Stretched Thin

Army Forces for Sustained Operations

Lynn E. Davis, J. Michael Polich, William M. Hix,
Michael D. Greenberg, Stephen D. Brady,
Ronald E. Sortor

Prepared for the United States Army

Approved for public release; distribution unlimited

ARROYO CENTER

The research described in this report was sponsored by the United States Army under Contract No. DASW01-01-C-0003.

Library of Congress Cataloging-in-Publication Data

Stretched thin : Army forces for sustained operations / Lynn E. Davis ... [et al.].
     p. cm.
     "MG-362."
     Includes bibliographical references.
     ISBN 0-8330-3816-8 (pbk.)
     1. United States—Armed Forces—Recruiting, enlistment, etc. 2. United States—
Armed Forces—Appropriations and expenditures. 3. United States—Armed
Forces—Personnel management. 4. United States—Armed Forces—Cost control.
     I. Davis, Lynn E. (Lynn Etheridge), 1943–

     UB323.S787 2005
     355'.033273—dc22
                                                                          2005016169

The RAND Corporation is a nonprofit research organization providing objective analysis and effective solutions that address the challenges facing the public and private sectors around the world. RAND's publications do not necessarily reflect the opinions of its research clients and sponsors.

RAND® is a registered trademark.

Cover photo courtesy of NASA

Published 2005 by the RAND Corporation
1776 Main Street, P.O. Box 2138, Santa Monica, CA 90407-2138
1200 South Hayes Street, Arlington, VA 22202-5050
201 North Craig Street, Suite 202, Pittsburgh, PA 15213-1516
RAND URL: http://www.rand.org/
To order RAND documents or to obtain additional information, contact
Distribution Services: Telephone: (310) 451-7002;
Fax: (310) 451-6915; Email: order@rand.org

Preface

The nation faces the critical challenges of having to provide military forces for sustained military operations abroad while protecting the American homeland. Our purpose is to describe an approach to measuring the availability of Army active-duty and reserve combat units by examining a range of potential operational requirements, force structures, and reserve policies. What emerges from our analysis is an understanding of the difficult trade-offs the Army faces and what this will mean for the future size, structure, and policies of active and reserve forces. This report will be of interest to anyone concerned with how the Army will be able to support sustained and worldwide operations both today and in the future.

In the Army, this research was sponsored by Major General David C. Ralston, Director of Force Management, Office of the Deputy Chief of Staff, G-3. It was conducted in RAND Arroyo Center's Manpower and Training Program. RAND Arroyo Center, part of the RAND Corporation, is a federally funded research and development center sponsored by the United States Army.

For more information on RAND Arroyo Center, contact the Director of Operations (telephone 310-393-0411, extension 6419; fax 310-451-6952; e-mail Marcy_Agmon@rand.org) or visit Arroyo's web site at http://www.rand.org/organization/ard/.

Contents

Figures

Tables

Summary

The Problem: Supporting Sustained Operational Deployments

Recent events have seen a growing demand for use of the nation's military forces, both for overseas operations and for homeland security. The increased operational tempo, particularly driven by the situation in Iraq and Afghanistan, has led to more frequent and lengthy deployments of units and soldiers across the entire U.S. Army. These in turn have placed increasing stress on the Army as it seeks to preserve its institutional commitments to training its soldiers and units and to maintaining a pool of ready units that can respond rapidly to new contingencies.

This situation confronts the nation with several key questions: Are the Army's active and reserve forces the right size to meet these demands? Does the Army have the right number and types of combat units to sustain high levels of overseas deployments while maintaining ready units for other possible contingencies? And how much does the rapid rotation of deployments stretch the Army's units and soldiers? The current report endeavors to address these questions and to examine alternative ways in which the Army might respond to current and future demands on its forces.

Effects of Deployments on Unit and Force Readiness

Our analytic strategy for addressing these questions focuses on large combat formations, or brigade combat teams (BCTs),[1] and involves examining a broad range of operational requirements. We stipulate a steady-state requirement for sustained deployments and compare that requirement with the supply of brigades that can be provided from the Army's active component (AC) and reserve component (RC) (Army National Guard), given alternative policies for utilizing the forces. From that analysis we derive two key outcome measures that describe critical aspects of the Army's ability to fulfill its missions: time at home[2] between deployments for AC BCTs and the number of "ready" AC BCTs. Unit time at home is important because it has wide ramifications for Army capabilities and the welfare of soldiers, including potentially recruitment and retention.[3] The number of ready units offers a metric for assessing the nation's defense posture and the Army's ability to respond rapidly to new contingencies and threats.

Those two outcome measures depend on several factors that may vary simultaneously:

- **Size of the operational requirement:** In our analysis, these ranged from 8 brigades to 20 brigades required for recurring overseas deployments at any given time.
- **Army force size and structure:** We examined both the baseline force (pre-2004) and the Army's planned transformed force

---

[1] A BCT typically includes a single maneuver brigade (such as an armor or mechanized infantry brigade) and various combat support and combat service support elements that deploy with it. The specifics vary across different types of brigades, but the nonmaneuver elements commonly include engineers, intelligence, military police, medical, transportation, and other support assets.

[2] For "time at home," the Army is using the term "dwell time." See Preston (2005).

[3] The unit's time at home between deployments is an important factor in determining the amount of time that individual soldiers can spend at home. However, an individual's experience over a career is also influenced by other factors, such as assignments to Korea and to institutional positions. These are examined in Chapter Three.

(expected to be complete by 2007 in the AC and 2010 in the RC).

• **Employment policies:** We analyzed variations in the duration of overseas deployments for both active and reserve forces, the frequency of mobilizations of RC units, and the amount of preparation time that RC units need before deploying.

The aim is to portray outcomes of various policy choices and thereby to assist policymakers in seeking to reduce the stress of sustained operations on the Army's combat forces.

Using the Active Component Alone

Our initial analysis focused on a base case in which the Army supports its operational demands exclusively by relying on brigades in the AC. As shown in Figure S.1, we defined four cases of operational requirements for recurring overseas deployments (8, 12, 16, and 20 brigades in theater at any time). We also specified the types of heavy, medium (Stryker), and infantry brigades, weighting them toward heavy-medium units, in line with the force mix the Army has sent to Iraq and Afghanistan. Figure S.1 portrays these four cases along with two cases of combat force supply: the baseline (pre-2004) Army force structure of 33 BCTs (32 rotating) and the posttransformation Army force of 43 (41 rotating) modularized brigades, called brigade combat team units of action (BCT-UAs).[4] The figure shows time at home separately for heavy-medium units (labeled H-M) and infantry units (IN).

What emerges from this analysis is that the baseline AC inventory of heavy-medium BCTs is placed under considerable stress when sustained deployment requirements exceed 10 brigades. At larger requirements (12 through 20 brigades), time at home for heavy-medium BCTs is less than two years—a well-established Army goal

---

[4] See Chapter Two and the appendix for a description of the Army baseline and transformed force structure. We will for simplicity use the term "transformed brigade" for the successor combat brigades in the Army transformation plan.

Figure S.1
AC Time at Home by Type of Combat Unit for Different Operational Requirement Levels

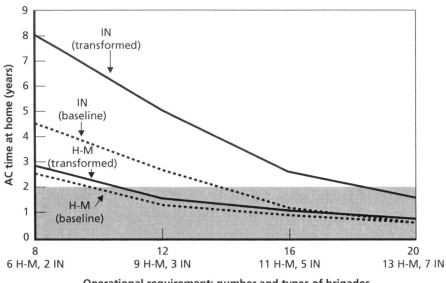

RAND MG362-S.1

for refreshing, refitting, and retraining its units between subsequent deployments. The situation for heavy-medium units is only mildly improved by the Army's plans for transformation.[5] Alternatively, some improvement can be made by permitting units of any type to substitute for one another (e.g., sending a light unit for a heavy requirement). However, as requirements rise above 14 brigades, all types of units have less than the goal of two years time at home. At these high levels of demands, the nation is left with few ready brigades for other potential contingencies.

---

[5] The goal of two years time at home is achieved for transformed infantry units until the requirement rises above 18 brigades. If, on the other hand, the rotation requirements shifted to emphasize infantry units, the strains on heavy-medium units would decrease. But infantry units would then be increasingly stressed.

Adding the Reserve Component to the Deployment Pool

The above results represent the situation if the Army uses only its AC units. What if it also calls on its RC units? In analyses to address that question, we examined the contribution of National Guard brigades to improving the time at home for AC brigades and the count of Army ready units. We focused on fulfilling a rotational requirement of 16 brigades–the size of the force actually deployed to Iraq and Afghanistan in mid-2004–specifying that 11 of these should be heavy or medium brigades.

The results show that AC time at home for transformed heavy-medium brigades is substantially less than two years, even assuming that the Army deploys its RC forces under a policy that mobilizes them one year out of every six years[6] and even after the posttransformation supply of AC and RC combat units is available for deployment. When we investigated a series of modified RC employment policies, such as more frequent RC mobilizations (e.g., one in four) or reduced RC preparation times (with longer RC deployments), we found these to be somewhat helpful, but insufficient even in combination to bring AC time at home for transformed heavy-medium brigades up to the two-year threshold.

Therefore, to meet a goal of two years AC time at home, the Army would need to take further steps beyond adding its transformed RC brigades to the rotation. One possible step would be to increase the supply of AC and/or RC transformed heavy-medium brigades beyond those currently in the transformation plan—a costly but conceivable solution. Another approach would be to permit flexibility by using different types of units to meet operational requirements. In that case, the nation would have to accept appreciable operational

---

[6] Secretary of Defense Donald Rumsfeld has issued guidelines that direct the services to plan for using RC forces on a schedule that results in only one year of mobilization out of every six. See a memorandum by the Secretary of Defense, July 2003, followed by a Department of Defense report on *Rebalancing Forces: Easing the Stress on the Guard and Reserve* (2004). The *Army Strategic Planning Guidance* of 2005, however, sets the goal in terms of "deployments": "one year deployed and five years at home station" for the National Guard, and "one year deployed and four years at home station" for the Army Reserve. This would require a frequency of RC "mobilization" of more than one year in six years.

risk, by assuming that any type of RC unit would be able to accomplish the mission and each of the National Guard brigades would be equal in capability to those in the AC.

Policy Options and Risks

Based on the above findings, we assessed policy alternatives that the Army could consider to manage intensive deployments and maintain readiness. What emerges is that each of these alternatives involves significant costs and risks. Therefore, we consider a series of future conditions that could emerge and pose questions about how the Army can adapt so that it can meet its immediate operational requirements and sustain its force over the longer term. If it cannot meet all needs under current plans, how could it adapt to improve the situation?

Suppose, Initially, That Overseas Rotation Requirements Drop Back to Ten Brigades. With that demand, and assuming that the Army both has the resources to implement its AC and RC transformation plans and can draw on all the National Guard brigades one year in every six years, all types of AC units would have at least two years at home between deployments. The Army would have more than 20 brigades ready for other contingencies, of which at least 11 would be heavy-medium units.

The issue for the nation is whether policymakers are comfortable basing future Army planning on this lower level of overseas rotational requirements. This assumption could be plausible if one views the current requirements in Iraq and Afghanistan as an aberration or something to be endured for a short time now or only periodically in the future.

Alternatively: What If High Overseas Rotation Requirements Continue for Some Time? To meet requirements levels in the upper range that we have considered—14 to 20 brigades—the Army would experience serious problems in AC unit readiness and the nation would have few if any ready AC brigades to turn to in a crisis. Transforming the Army into the planned structure of 43 active transformed brigades will help, but transformation is largely in the future, comes with its own uncertainties, and cannot meet the full demand for rotational forces by itself.

The nation could decide to live with these low levels of ready AC units and training time—if it assumed that the Army will only rarely need to respond quickly to contingencies with large numbers of forces either overseas or at home. In effect, this course means assuming that international or domestic contingencies will not require Army combat brigades to do much beyond supporting overseas rotations.

What If the Risks Are Too High for the Army to Plan for Low Levels of Contingency Requirements? As we have described, there are two possible adaptations. The Army could turn to the RC and plan on utilizing them at reasonable rates—e.g., mobilizing all National Guard brigades for one year out of every six years. However, these units can be called only at reasonable intervals, and they can cover only a modest portion of the requirement for overseas forces, even assuming, as the Army does, that all transformed National Guard brigades will be capable of participating in the rotations. Alternatively, the Army could plan to fill rotational requirements based on the assumption that any unit could fulfill the mission. Such flexibility greatly improves the situation, but only if the transformed National Guard brigades are all available to be mobilized one year in every six years and all equally capable of meeting the overseas requirements. Such a course carries operational risk if the theater environment is not benign or missions require armor protection and on-the-ground mobility. To date, the Army has hedged against such risks by deploying forces to Iraq that are predominantly heavy. Moreover, when overseas rotation requirements increase beyond about 17 brigades, AC time at home falls below two years even assuming such flexibility.[7]

---

[7] Alternatively, to meet a 20-brigade requirement, some Army planners suggest accepting full flexibility *and* deploying RC brigades for a one-year tour every six years. That would achieve the goal of two years time at home for AC units, but it would require the Army to mobilize the RC brigades for 18 months in every six-year period (equivalent to a mobilization frequency of one in four years).

What If It Is Too Risky to Assume That Infantry, Medium, and Heavy Units in the AC and RC Can Substitute for One Another in Future Missions? We have explored two options to respond under those circumstances. One avenue is for the Army to forgo its transformation plans to convert heavy National Guard units to infantry units. This would also require the Army to find the resources to make all these units—including the divisional brigades—equal in readiness to AC brigades. Alternatively, the Army could take an approach that pursues its National Guard transformation plans and keeps RC utilization within current policy constraints but adds heavy force structure to the AC. This could be accomplished either by changing the mix of the units planned in the Army's transformation or adding additional transformed brigades. But this would call for finding billions of dollars well beyond the current Army modularity plan and would take years to achieve.

To decide on an overall approach for the future will require the nation to confront a number of trade-offs in terms of the Army's reliance on the AC and RC, in terms of the risks it is willing to take with the Army's ability to meet different types of future contingencies, in terms of what types of training of Army units will be required for different types of operations, and in terms of what resources are available for transforming the RC and increasing AC force structure. Our analysis suggests that the challenge is profound and that making the trade-offs will not be easy.

Acknowledgments

This report benefited from the support and assistance of many people in the Army and at RAND. In particular, we thank our sponsor Major General David C. Ralston and those on his staff who provided us with good counsel and critical information on the Army's goals and plans. We want especially to thank our RAND colleagues David Mosher and Brian Reid, who produced estimates of the costs of the policy options that we developed in the course of our analysis. We also appreciate the contribution of Jason Castillo, Andy Hoehn, David Kassing, Chip Leonard, Tom McNaugher, Jim Quinlivan, and Lauri Zeman. They provided comments on our analysis all along the way. Special thanks goes to Bruce Orvis, the Director of the Manpower and Training Program, whose careful reading of our drafts has resulted in many improvements. We appreciated the thoughtful and insightful reviews of the first draft of our report by Michele Flournoy, Bryan Hallmark, and Colonel Paul Thornton. Dan Sheehan has been our editor again, showing his skills with language and his understanding of the authors' many wishes. Thanks to Dan and to all in the RAND Publications Department who helped in the production of this report. The content and conclusions of the report, however, remain solely the responsibility of the authors.

Abbreviations

AC	Active component
BCT	Brigade combat team
CBO	Congressional Budget Office
CTC	Combat Training Center
DoD	Department of Defense
GAO	Government Accountability Office
H-M	Heavy-medium
IN	Infantry
OMB	Office of Management and Budget
OSD	Office of the Secretary of Defense
RC	Reserve component
SBCT	Stryker Brigade Combat Team
TDA	Table of Distribution and Allowances
TOE	Table of Organization and Equipment
UA	Unit of action

CHAPTER ONE

Introduction

The Problem: Intensive and Frequent Operational Deployments

Recent events have seen a growing demand for use of the nation's military forces, both for overseas operations and homeland security. The increased pace, driven by the situation in Iraq and Afghanistan, has led to more frequent and lengthy deployments of units and soldiers across the entire U.S. Army. In those operations, units are deployed to the theater for an extended period (usually, one year or longer), replacing an existing unit and in turn being replaced when it returns to its home station. The resulting rotation pattern means that much of a unit's time is devoted to deployments or to recovery from a previous deployment and preparation for the next one. Because the scale of operations is very large—at this writing upward of 16 Army brigades are deployed in Iraq and Afghanistan—the effects reverberate throughout the force.

With these new and very demanding calls on Army forces, the nation faces some novel questions. Are the Army's active and reserve forces the right size to meet these demands? In particular, does the Army have the right number and types of combat units to sustain high levels of overseas deployments while still maintaining ready units for other possible contingencies? Still more specifically, but highly germane to the debate: How much does the rapid rotation cycle of deployments stretch the Army's units and soldiers? This report addresses these questions and analyzes alternative ways in which the Army might respond.

The Effects of Intensive Deployments

Intensive deployments lead to two key problems that the Army needs to manage: effects on units and effects on individuals. For units, rotational deployments reduce the number of units ready at home for contingencies that could arise quickly. They also disrupt other unit training activities, particularly cycles of unit training in which combat units develop their skills and collective capabilities to prosecute major conflicts. These cycles typically begin with small-unit exercises and culminate in large force-on-force exercises at home station or combat training centers. In an environment dominated by repeated overseas deployments, units are often unable to complete such training. Failing to accomplish wartime training also affects the development of leaders in the enlisted and officer corps.[1] Finally, to the extent that units experience personnel turnover after a deployment, the unit is no longer ready for another rotation similar to the one they just returned from. Deployments thus reduce the readiness of the force to meet all types of emerging contingencies.

Deployments also create turbulence for individual personnel, with several important ramifications. First, deployments take soldiers away from their homes and communities, thus reducing quality of life for both soldiers and their families. If sustained over the long term, this may reduce morale and hinder the sustainability of manpower levels by lowering retention rates. Second, deployments impose additional workload for preparation, planning, and maintaining units that are on the move. For example, staff planners and support functions are stressed by "split-base" operations in which part of the unit is overseas while another part is still at home station. Third, turbulence inevitably causes some soldiers to be away from the unit (and from its

---

[1] Many observers believe that enlisted soldiers and junior officers in small units in Iraq gain leadership skills and experience that are notably better than they might gain in most training venues. However, except in very rare cases, they and the more senior leaders are not getting the experience or developing skills in coordinating the movement and fires of their armor, infantry, artillery, and helicopter assets or the fires of Navy or Air Force elements. These skills are gained over multiple assignments and training experiences with practice against similarly well equipped and trained adversaries. Not only must the leader have these skills to lead his unit in a combat environment, but also—perhaps more importantly—he needs the experience and skills so that he may train subordinates.

home-based training facilities) while the unit is conducting collective training. Thus, some training is accomplished without all the soldiers who will eventually be needed for in-theater operations.

Using the Active and Reserve Components

These problems affect both the active and reserve components, and they show no signs of abating. In fact, trade-offs between using the active component (AC) and the reserve component (RC) have figured prominently in recent public debate about the nation's military posture. Especially for initial phases of operations, the Army often prefers to use the AC.[2] Over time, however, this heavy utilization of AC units creates problems. The more the AC is used, the less time AC units have at home for recovery and training, and the fewer ready units will be available for other missions. To ameliorate the problems, the Army can turn (and has turned) to its RC, but similar problems emerge there, and new problems are added. For example, to be deployed overseas soldiers in the National Guard must be mobilized and afforded special training, which can be lengthy. The preparation period adds to the time that National Guard soldiers are away from their homes and civilian jobs. For all these reasons, the Department of Defense (DoD) has sought to limit intense utilization of the RC, articulating a policy that seeks to limit reserve mobilizations to no more than one year in six years over the long term.[3]

---

[2] See, for example, a memorandum from Secretary of Defense Donald Rumsfeld setting goals for structuring active and reserve forces to "eliminate the need for involuntary mobilization [of reserves] during the first 15 days of rapid response operations" (Rumsfeld, 2003). At the same time, the Army is placing more combat support and combat service support units into the active Army to improve its deployability and ability to sustain operations during the first 30 days of a campaign (Department of the Army, 2005b).

[3] In the same memorandum, the Secretary of Defense set goals for the frequency of RC mobilizations for deployments: "structure forces in order to limit involuntary mobilization to not more than one year every six years" (Rumsfeld, 2003). The *Army Strategic Planning Guidance* (Department of the Army, 2005c), however, sets goals in terms of "deployments": "one year deployed and five years at home station" for the National Guard, and for the Army Reserve, "one year deployed and four years at home station." Details of these goals and limitations will be discussed in Chapter Three.

Thus, difficulties, costs, and downsides are inherent in using either the AC or RC too intensively to support deployments. The problems have come into sharp focus as recent operations led to positioning large numbers of units amounting to more than 150,000 personnel in Iraq and Afghanistan. Moreover, it seems prudent to assume that the level and pace of activity will remain high for some time, perhaps years. This leads to a central policy question, which is the subject of this report:

- How can the nation use both the AC and RC to meet future overseas rotational demands over the long term and provide sufficient ready units for other operational contingencies, while maintaining adequate training opportunities for units and keeping individual deployment times to reasonable levels?

Analysis Strategy

To address the above question, we developed an analysis strategy that was guided by the experience of the recent past but also flexible enough to let us vary several key parameters to see how results would change. Our strategy considered three types of conditions that might change, as depicted in Figure 1.1: the scale and nature of operational requirements for overseas rotations; the Army active and reserve force structure; and policies governing employment of active and reserve forces.

As shown in Figure 1.1, the analysis will consider various changes in these conditions and then will assess how those changes would affect two key outcomes: the amount of time that units have at home between deployments[4] and the number of ready units the Army has available at any time to serve other national purposes. Below we explain the analysis strategy and the types of variations that we will treat.

---

[4] For "time at home," the Army is using the term "dwell time." See Preston (2005).

Figure 1.1
Analysis Strategy

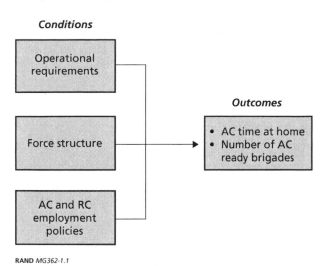

RAND *MG362-1.1*

Operational Requirements

All discussions of future military requirements are shrouded in uncertainty. Guidance from DoD seeks to move military planning from traditional threats (preparing for conventional wars) to what are termed the "strategic challenges" presented by irregular threats (terrorism, insurgency), catastrophic threats (attacks with WMD), and disruptive threats (breakthrough technologies in the hands of enemies) (*The National Security Strategy*, 2002, p. 4). Translating these strategic challenges into operational requirements is just under way within the Defense Department. The most recent defense planning guidance has called for sizing U.S. military forces based on a construct described as "1-4-2-1." This guidance calls for the ability to accomplish these missions: defend the U.S. homeland and territory against external attack; deter aggression and coercion in four critical regions: northeast Asia, the east Asian littoral, the Middle East and southwest Asia, and Europe; swiftly defeat the efforts of an adversary in two overlapping wars while preserving the option to seek decisive victory in one of those conflicts; and conduct a limited number of lesser contingency operations.[5] Whether this construct remains appropriate in the aftermath of the September 2001 terrorist attacks and the war in Iraq is a matter of debate, introducing another uncertainty in future force planning.[6] The Chief of Staff of the Army in early 2005 described his planning requirement: "to be able to deploy and sustain 20 brigade combat teams" (U.S. Senate, 2005).[7]

In the course of our analysis, we will recognize all these uncertainties, but we begin by varying the magnitude and characteristics of future Army requirements for recurring overseas deployments. We will examine cases in which the number of brigades deployed overseas at any one time ranges from a low of 8 brigades to a high of 20 brigades. We will also consider the *types of units* that may be required in these deployments, such as heavy units (containing armor or mechanized elements), Stryker brigades (more-mobile elements based largely on wheeled vehicles), and infantry units.

We recognize that requirements for recurring overseas deployments are only a part of the defense planning space. For example, additional antiterrorism operations could arise. Hostile nation-states could undertake aggression that the United States would need to deter or repel. Homeland defense and homeland security operations could impose greater demands for Army forces. In actual execution, however, the immediacy and urgency of recent overseas rotations have meant that their requirements had to be satisfied first. In effect, forces were sent to Afghanistan and Iraq even though their readiness and availability for some other missions was affected.

---

[5] See Department of the Army (2003a, Annex A, pp. 16–17) and Joint Chiefs of Staff (2004).

[6] According to the Department of the Army (2005a), "although not specifically enumerated, capabilities and force structure for stability operations and for the war on terrorism are now included in the construct as elements that span the entire range of activities described in the construct."

[7] This 20-brigade requirement appears to include requirements for recurring rotational deployments, forward stationing of brigades in Korea and Europe, as well as brigades for other contingencies, including major conflict operations. According to the Secretary of the Army, Francis J. Harvey, the Army should also have the capability "to surge and provide more" (Harvey, 2005).

Our analysis follows the same path: We will posit the size of overseas rotational deployment requirements and then determine what forces remain in a sufficient state of readiness to be useful for other purposes, whether they be in Korea or the Middle East, the war on terrorism, homeland defense and security, or other combat or stability operations.

Force Structure

A second major uncertainty in the future policy environment involves the supply of Army units available for overseas deployments. This supply is changing as the Army undergoes a transformation from a combat structure based on divisions and brigades into a new structure that places more emphasis on brigade-size elements, which are expected to be more numerous after transformation is complete.

This report focuses on requirements for large combat formations, specifically brigade combat teams (BCTs).[8] The initial deployments to Iraq and Afghanistan came from an active Army that contained 33 such brigades. This we term the "baseline force," that is, the force that existed as of early 2004.

During 2004, however, the Army began to transform its overall combat structure in a way that is intended to create a more modular and flexible force. This transformation will reduce the size of divisional elements, move some of those elements into the BCTs, and create more brigades, called brigade combat teams (unit of action) (BCT [UA]), by the Army.[9] The current plan calls for creating at least 43 active transformed brigades in place of the 33 BCTs that pre-

---

[8] A BCT typically includes a single maneuver brigade (such as an armored or mechanized infantry brigade) and various combat support and combat service support elements that deploy with it. The specifics vary across different types of brigades, but the nonmaneuver elements commonly include engineers, intelligence, military police, medical, transportation, and other support assets.

[9] See Department of the Army (2005c). Official terminology for these new entities has been in flux. We are focused in this report only on the current BCTs and their successor brigades, whatever they might be called in the future. For simplicity in our discussion, we will use the term "transformed brigade" and mean only the BCT (UA) and not any of the other UAs being created as part of the Army transformation.

viously existed. To reflect this potential growth in assets, our analysis will portray results based on two cases: using the *baseline force* composed of traditional BCTs and using the *transformed force* composed of transformed brigades.

The transformation of AC forces is already under way and is expected to produce ten new transformed brigades by the end of 2006, with the full conversion of all BCTs to transformed brigades due in 2007.[10] The transformed brigade, according to the Army, is a "stand-alone and standardized tactical force of between 3,500 and 4,000 Soldiers."[11] The National Guard is also expected to transform but in a somewhat different way. As we will discuss in Chapter Three, the RC transformation will not result in larger numbers of transformed brigades, but it will change the mix of heavy, medium, and infantry units. As we consider the potential role of the National Guard, our analyses will reflect the plans for RC transformation. The Army is also planning to resource 12 Army expeditionary packages to provide an efficient and continuous supply of combat and support soldiers for combatant commanders while providing predictable unit deployment schedules for soldiers and their families (Department of the Army, 2005c).[12]

In our analysis, we have accepted the expectation that the Army will eventually have 43 transformed brigades and that each will be capable of performing the same tasks as a previous BCT. We will also assume that the personnel strength of the transformed brigades will be roughly equal to the BCTs and that the resources will be available to create and sustain those manning levels and capabilities.[13] How-

---

[10] At about that time, the Army expects to make a decision about whether to continue transformation to attain an eventual structure that incorporates 48 transformed brigades. However, present plans call for only 43 transformed brigades; further expansion is said to depend on operational necessity and approval by senior DoD officials. See Department of the Army (2005a; 2005b).

[11] See Department of the Army (2005c).

[12] As of this writing, the Army is fleshing out this concept.

[13] The Department of Defense added $35 billion over seven years (FY 2005–2011) to the $13 billion in the Army baseline budget for a total of $48 billion for Army "modularity." About $10 billion of this funding in FY 2005 and FY 2006 will be in supplemental appro-

ever, for several reasons it may be prudent to anticipate a situation that falls somewhere between the results that we portray for the baseline and transformed forces, at least in the near term. For one thing, the interim level of transformation—attaining a fully equipped set of 43 transformed brigades—is scheduled for 2007. If delays are encountered in the meantime, fewer transformed brigades will be available even after that date is reached. In addition, many specifics of the transformation plan are in flux, particularly the size and composition of support elements that may be needed to deploy with the combat brigades. The eventual capability of the transformed forces may depend on how these elements are organized and, importantly, whether sufficient resources (manpower and funds) are available to constitute them. Finally, the resource issue, if it becomes a limitation, may in turn limit the number of maneuver transformed brigades that the Army can afford.[14]

AC and RC Employment Policies

In addition to changes in requirements and structure, the Army has goals for using the various components, but these too the Army may wish to alter. We will examine three types of policies that exert profound effects on the readiness and utilization of deploying units: the *duration of active and reserve overseas tours;* the *frequency of mobilization* for RC units; and the *amount of preparation time* that RC units need before deployment.

---

priations. These funds will cover procurement of equipment plus additional facilities and infrastructure. See *President Bush's FY 2006 Defense* Budget (2005) and OMB (2005).

[14] Some resource uncertainties are apparent even in the current situation, in which extraordinary adjustments have been made to allow for the strains of current operations. For example, the Army's estimate for modular transformation has increased substantially over the past year. Funding for the next two years of the Army's modular transformation will be drawn in part from contingent and supplemental appropriations. The official Defense Department plan is to increase Army end strength by 30,000 temporarily and then return to its pre-2001 level in 2009, having increased the pool of usable and deployable troops by drawing existing soldiers from other parts of the Army and by converting some uniformed military positions to civilian positions. See OSD (2005) and Department of the Army (2005b). For a discussion of these and other uncertainties, see Feickert (2004) and Pickup and St. Laurent (2005).

Each of these policies poses its own set of choices and trade-offs. For example, in recent years the Army has planned to deploy AC and RC units for one year—that is, a unit spends a continuous 12-month period in the overseas theater performing an operational mission before returning to its home station. In practice, some AC deployments have been extended to meet operational exigencies.

For the combatant commander, longer tours reduce turbulence and increase the experience of units in the theater. Congressional and other observers have sometimes urged that all deployments be shorter, perhaps lasting only six months.[15] We will analyze the effects of varying both AC and RC deployment durations.

Another key policy is the frequency with which RC units are mobilized. DoD has stated as a planning goal its desire to mobilize RC units in such a way that they spend, on average, only one year mobilized out of every six years. Again, in practice this has proven difficult to achieve for some types of units, and we will examine how variations in this parameter affect outcomes.

Finally, an important limitation on using RC forces is the amount of preparation, training, and recovery time the units require when they are mobilized for a deployment. Most RC brigades have required about six months to prepare for deployment to Iraq.[16] Because any time for these "overhead functions" reduces the time they are available to serve in theater, we will analyze the effects of reducing preparation time to determine whether new policies involving training and additional resources aimed in that direction would yield substantial benefits.

Assessing Outcomes

Our analysis will evaluate variations in the three conditions described above in terms of two key metrics: *AC time at home between deploy-*

---

[15] In fact, the former acting Secretary of the Army reportedly urged planning for shorter tour lengths as an option for the future (Shanker, 2004, p. 23).

[16] For RC brigades mobilized through the end of 2004, the period from mobilization to deployment has ranged from four to seven months, yielding an average of about 5.8 months, according to Associated Press wire service reports.

ments and *the number of ready AC brigades* available for other national needs.

Why focus on these two things? The first metric, time at home between deployments, has wide ramifications for Army capabilities and for the welfare of soldiers. If units rotate too quickly between deployments, they lack the time at home to train and develop readiness for any other contingencies. The amount of time that a soldier is able to spend at home station—with his or her family and near training and support structures on base—is also a fundamental measure of soldiers' quality of life, a criterion that has figured prominently in public debate for at least a decade.[17] The unit's time at home between deployments is an important factor in determining the amount of time that individual soldiers can spend at home. However, an individual's experience over a career is also influenced by other factors, such as assignments to Korea and to institutional positions, which we will examine in Chapter Three. While we have no conclusive evidence about the long-term effect of sustained deployments on recruitment and retention, it seems likely that they are linked to time at home.[18]

The second metric, number of ready units, stems from the first metric but represents a different way of assessing the nation's defense posture. When trouble brews on the international scene, the President needs ready units that can be tapped. If units are so busy rotating back and forth to deployments that they cannot train for war, the number of truly ready units may drop to a small number. In fact, as

----

[17] For example, just a few years ago Congress enacted provisions that specified ceilings for the amount of time that any individual military member should spend away from home, enjoined the services to meet those ceilings, and required substantial financial compensation to soldiers whose time away from home rose above the ceiling (U.S. Congress, 1999).

[18] According to General Richard Cody (2005), "We would like to go to a 24 month, one year in, two years out for the [AC] force [i.e., one year deployed, two years at home]. We think that's what we can sustain an all-volunteer force for." See also a statement released by the Government Accountability Office (GAO) (2005). DoD policy changes involving utilization of the RC have created many uncertainties concerning "the likelihood of their mobilization, the length of service commitments and overseas rotations, and the types of missions they will have to perform. The uncertainties may affect future retention and recruiting efforts, and indications show that some parts of the force may already be stressed."

we shall see, there are plausible cases in which the Army could have *no* ready units. We therefore will portray this measure of time at home as a single, convenient measure of force readiness.

An Evolving System

Our analysis will focus primarily on the above variations and policy measures. However, it should be noted that the modern Army is a rapidly changing institution with numerous new policies and initiatives under way. Two of these merit mention here because they are closely related to our topic. First is an ongoing initiative to "rebalance" the Army force structure, whereby personnel spaces for lower-priority structure (e.g., field artillery and air defense specialties) are converted to higher-priority AC and RC structure (e.g., chemical, military police, engineer, medical, quartermaster, and transportation specialties) (DoD, 2004a). These steps do not directly affect our analyses of combat brigades, but they would be very important for analysis of the Army's support forces.

A second area of change is in unit manning. In recent years, the Army has implemented new policies to man operational units at 100 percent, rather than at the lower levels that had previously been permitted. Beyond that, the Army envisions "life-cycle manning" of units, in which units would periodically be re-formed with an entirely new group of soldiers and then kept together for several years. The object is to encourage unit cohesion and reduce individual personnel turbulence that naturally grows out of the Army's traditional system of "individual replacements." A life-cycle manning system, if implemented across the Army, could affect the availability of brigades for overseas deployments, which we will consider briefly.[19]

Organization of This Report

The remainder of this report deals with four main topics. Chapter Two describes how the Army could support various operational requirements for deployments if it used only the AC. In the course of

---

[19] See Army Campaign Plan briefing, available at http://www.army.mil/thewayahead/acppresentations/4_12.html.

that analysis, we explain our basic methodology for calculating time at home and the number of available AC brigades, and we also show how challenging it is to support current levels of overseas operations.

Chapter Three describes the situation when RC forces are employed along with the AC. It will show how much the RC can contribute and will analyze how various changes in employment policy could ameliorate the situation. It concludes with a brief analysis of the effects of sustained overseas rotations on individual soldiers and the amount of time they spend away from home during their career.

Chapter Four describes different approaches to planning for a future that involves high levels of Army overseas rotations. It defines seven options to assess trade-offs involving the number of AC units and the degree of dependence on the National Guard.

Chapter Five presents our conclusions. It aims to place our results in a broad policy context and to articulate the choices that the nation faces in supporting continual overseas rotations of its military forces both today and in the future.

The appendix provides backup documentation on Army force structure and the methods used to calculate AC time at home, RC utilization rates, number of ready AC brigades, individual soldier time away from home, and our cost estimates.

CHAPTER TWO

Employing the Active Component

This chapter defines the types of active Army brigade-level elements considered in the analysis and then describes the range of future operational requirements that we posited. Based on those assumptions, it then assesses the effects when all of these deployments are satisfied using AC units.

Force Structure Available

As described in the introduction, the baseline Army (pre-2004) contained three major types of ground maneuver brigades:

- heavy brigades, generally armor or mechanized infantry;
- medium (Stryker) brigades, known as SBCTs (Stryker Brigade Combat Teams); and
- infantry brigades, an Army classification that refers to light and airborne brigades, which typically lack an extensive complement of ground combat vehicles.

Table 2.1 shows counts of AC brigades according to those three categories.[1] The Army has used this classification in its transforma-

---

[1] See the appendix for a listing of the brigades in the baseline that fit into these categories. In our baseline force, we include in the Stryker category the 2nd Armored Cavalry Regiment because of its wheeled inventory and mission capability, even though it had not yet converted.

tion and equipment planning, and so we adopted it for our analysis. The classification is important because the different types of units are specialized for particular combat missions and environments, and there are limitations on the extent to which they can readily substitute for one another. The predominant portion of the inventory is heavy, although the Army in 2004 had already outfitted new Stryker brigades to provide a more mobile and versatile combat capability. Their equipment and capabilities provide ground mobility but less overall protection than heavily armored vehicles.

Army brigades have other things to do besides standing ready for potential deployments, of course. For example, in addition to sustaining their warfighting skills, the brigades in Korea have a deterrence mission. Although the Army recently decided to use one brigade in Korea as part of the rotation plan for Iraq, we concluded that it would be reasonable to retain at least one other brigade in Korea permanently (that is, not participating in the rotational schedule). Therefore, the second column of Table 2.1 shows that the AC baseline force really only has 32 brigades available for recurring overseas rotations.

Somewhat different characteristics apply to the planned future force, after transformation, as shown in Table 2.2. Most existing divisions that contain three ground maneuver brigades will be reorganized to create four brigade-like units. These new transformed brigades are expected to contain roughly the same combat power and about the same number of soldiers as existing BCTs. Altogether the plan

Table 2.1
AC Force Structure: Baseline

Unit Type	AC Forces (Number of Brigades)	
	Inventory	Rotating[a]
Heavy	18	17
Medium (SBCT)	4	4
Infantry	11	11
Total	33	32

[a] Not rotating: one heavy BCT in Korea.

Table 2.2
AC Force Structure: After Transformation

Unit Type	AC Forces (Number of Brigades)	
	Inventory	Rotating[a]
Heavy	20	19
Medium (SBCT)	5	4
Infantry	18	18
Total	43	41

[a] Not rotating: one heavy transformed brigade in Korea and one SBCT in Europe.

calls for 43 transformed brigades to replace the existing 33 BCTs. Most of the growth will be in the number of infantry units (Department of the Army, 2004b).[2]

For our analysis, the important facts are shown in the number of units available for overseas rotations. Considerable uncertainty exists about whether all these will be available. For primarily political reasons, we consider as forward stationed, and therefore not available, one heavy transformed brigade in Korea and one SBCT in Europe. We will also assume that the line between heavy and Stryker units is indistinct: Some Stryker brigades may be used to fill nominal requirements for heavy units, and some heavy units may substitute for Strykers. So we will combine the heavy and Stryker units into a category called "heavy-medium" units. However, we will generally not assume that infantry units can fill requirements calling for heavy or Stryker units. Thus, to support rotations the Army would have the following:

- In the baseline active force: 21 heavy-medium units and 11 infantry units.
- In the transformed force: 23 heavy-medium units and 18 infantry units.

---

[2] As noted earlier, Army long-term plans contemplate further expansion to a total of 48 transformed brigades, but that decision will be made in the future.

Operational Requirements: Overseas Rotations

This analysis begins by defining the Army's operational requirements for overseas rotations. These involve recurring deployments, such as those today in Iraq, Afghanistan, and the Balkans. It will then assess the availability of Army combat brigades at home for other contingencies that could arise—for example, in the war on terrorism, in a major conflict operation, or in defense of the homeland. (The potential requirements for these "ready brigades" are considered later in this chapter.)

What the Army requirements for recurring overseas rotations will be in the future is very uncertain. In the not-so-distant past, the United States had only about four brigades deployed in total overseas, during the late-1990s operations in the Balkans. However, since the September 2001 terrorist attacks and the operations in Iraq, it seems unlikely that the Army can plan to return to that relatively low level of rotations anytime soon. Overseas rotation demands could stem from several sources. While not specifying requirements for overseas stability operations, the Defense Department has defined four cases where U.S. military forces could be involved. These include requests from friendly states for assistance in protecting themselves from subversion, lawlessness, and insurgency; operations that precede, are concurrent with, and follow major conventional combat operations; intervention in a nation or region that becomes ungovernable, collapses economically, and disintegrates into anarchy; and operations to defeat groups whose ideology involves significant degradation of human rights and actions to destabilize legitimate governments.[3]

Recent history suggests that these requirements can be large. The Army's requirements for operations in Iraq and Afghanistan have grown over the past year to almost 20 brigades. The Army today is still rotating units to Kosovo and the Sinai peninsula. In the future, the Army might be called on to replace coalition or U.S. Marine units in Iraq or Afghanistan, and most likely if requirements in these countries decrease, the Marines and possibly the coalition forces will be

---

[3] For a description of these cases, see DoD (2004b).

withdrawn before the Army. There could also be new calls on Army units in efforts to bring peace to the Middle East.

Because of these uncertainties, we decided to bound the range of future possibilities for recurring overseas rotations by defining four cases, starting with an eight-brigade requirement and ending with a 20-brigade requirement. See Table 2.3.

For each of these overall brigade requirements, we also stipulate a mix of heavy-medium and infantry forces, weighted toward the heavy-medium side. This reflects recent sourcing decisions in Iraq that tended to emphasize units combining mobility and some degree of armored protection. Because of future uncertainties, we show what happens if all types of units are deemed substitutable for each other.[4]

The 12-brigade requirement is exemplified by the level and types of forces that the United States planned to deploy to Iraq and Afghanistan in spring 2004.[5] That deployment included nine heavy-medium and three infantry brigades (counting both AC and RC Army units, but not Marine or international units). The 16-brigade requirement represents the level of forces and types close to the level that the Army eventually deployed for those two operations in summer 2004 (Schwartz, 2004). According to Army officials, the heavy units initially deployed "one-third heavy"—that is with one-third of

Table 2.3
Overseas Rotation Cases: Number of Brigades Required

Total	Heavy-Medium	Infantry
8	6	2
12	9	3
16	11	5
20	13	7

---

[4] Specifying types of units and then permitting flexibility is an analytical device. When deployment requirements are sourced, the readiness of existing units—essentially a "supply" consideration—is assessed, in addition to the combatant commander's preferences—essentially the "demand" for forces. So requirements are not truly flexible or inflexible.

[5] For a description of those planned deployments, see Department of the Army (2003b).

their tanks and two-thirds of the brigade with "up-armored" wheeled vehicles. However, the security situation changed, and so about 150 tanks and 100 Bradleys were sent back to Iraq to give them more armored capability. In planning for the return of the 3rd Infantry Division, the combatant commander requested heavier units (Cody, 2004; 2005). The 2005 rotation to Iraq is now planned to have 12 heavy-medium units and two infantry units, and the rotation to Afghanistan two infantry brigades. Two light National Guard brigades were in the first deployment to Iraq, but eight of the next ten National Guard brigades were heavy units. Given the pool of AC units available for the follow-on rotation in 2006, the force deploying will be more balanced in types of units (DoD, 2004c). Thus, the Army has been tailoring its BCTs by shifting emphasis between heaviness and mobility depending on the situation in the theater and the availability of units at home, but it has shown a general preference for having a preponderance of heavy-medium units in the theater.

Results

Based on the above specifications of the *demand* for rotating units (the operational requirements) and the *supply* of units (force structure), we now examine the results.

Effects on AC Time at Home

Baseline Force. Figure 2.1 shows AC unit time at home in years when deploying the baseline force, for each of the cases of overseas rotation requirements. Along the horizontal axis we show the four levels of operational requirements: 8, 12, 16, and 20 brigades overseas, with the specified types of units. The lower, dotted line represents heavy-medium BCTs (i.e., the combination of heavy and Stryker brigades, labeled H-M). The upper, heavy line represents infantry BCTs (labeled IN). Evidently, heavy-medium units are the most stressed.

To illustrate, let us outline the calculations underlying the case that calls for 16 brigades required overseas: 11 heavy-medium and five infantry. To meet the requirement for 11 heavy or medium

BCTs, the baseline Army has a total of 21 heavy-medium BCTs that could rotate (see Table 2.1). Therefore, a full cycle of rotation (i.e., a cycle in which consecutive deployments result in using all available force elements) lasts 1.91 years (21/11). Because each heavy-medium unit spends one year overseas within that cycle, its amount of time at home is .91 years, or about 11 months. (For details and generalization of this logic and corresponding formulas, see the appendix.)

By all accounts, 11 months at home is not a very long time.[6] On that cycle, each AC unit would be quickly alternating back and forth between its home station and the overseas theater, spending more time overseas than at home. This corresponds to what the 2004 situa-

Figure 2.1
AC Time at Home Using Baseline AC BCTs

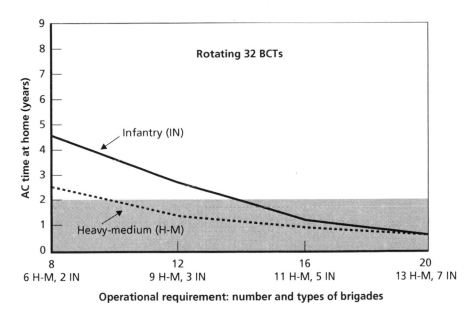

RAND *MG362-2.1*

---

[6] Defense officials have said that planners strive to maximize the amount of time spent at home station and to ensure that it exceeds the time deployed, even under the current extraordinary demands (Chu, 2004).

tion in Iraq and Afghanistan would have required for heavy-medium units, if only AC (and no RC) units were involved.

The situation would get worse, as shown by the rightmost point on the H-M line, if the requirement were to rise to 20 brigades. Then AC time at home would drop to .62 years, or seven months. On the other hand, if requirements were to relax, AC time at home would rise, reaching 2.5 years if only eight brigades were required.

We have judged these outcomes by stipulating a goal of two years at home for every year deployed. This represents a situation in which the unit spends about one-third of its time deployed and the rest at home. There is considerable precedent for such a goal, across all of the military services.[7] Thus, in Figure 2.1 we shaded the region under two years in gray. Whenever the time at home line drops into this "red zone," we argue that the situation has become serious and the force will be under considerable stress. In Figure 2.1, all deployment levels requiring more than eight heavy-medium brigades place the Army's heavy-medium forces within that "red zone."

The situation for the infantry forces, illustrated by the upper line in Figure 2.1, is more favorable. At a requirement of 12 brigades, the infantry forces reach a time at home of 2.67 years—although at higher requirements, they too drop into the shaded zone. This occurs because the requirements we have specified are weighted toward the heavy-medium forces. For example, at a requirement of 12, the infantry's portion of the deployment is only three brigades and they have 11 brigades available for rotation. This allows a slower pace of rotation and therefore more time at home.

To summarize: The goal of two years time at home is met for heavy-medium units at levels of overseas deployments of up to about

---

[7] The Army has clearly stated as its rotation goal to deploy AC units only one year in three years. See Department of the Army (2005c) and Department of the Army (2004a, Annex F). Similarly, the U.S. Navy has long operated on a schedule that keeps ships at sea for six months and in port for one year (yielding a ratio of three units in the structure for each unit deployed). During recent Quadrennial Defense Reviews, the services argued for deployment ratios of "three to one" or even "five to one" to account for the amount of time required to perform essential training, rest military personnel, refit equipment, and so forth. (See, for example, Department of Defense, 2001, and Congressional Budget Office, 2003.)

10 brigades. Above that, they slip farther and farther below the goal. Because of the larger number of infantry units in the Army force mix relative to their requirement in the rotation, the goal of two years time at home is achieved for infantry units at levels up to 14 brigades. These outcomes show why the Iraq and Afghanistan operations in 2004 have placed the active Army under such strain and why the Army quickly turned to its RC forces as a supplement.

Transformed Force. After transforming to 43 transformed brigades, the AC will have more brigade-sized units available. How much will this help? Figure 2.2 shows the result. The results for the transformed force are identified as the "41 case" (41 rotating transformed brigades). The upper heavy line represents infantry units, and the lower heavy line represents heavy-medium units. To facilitate comparisons between the two cases, Figure 2.2 retains the lines (dotted here) for the baseline force (identified as the "32 case") as shown in Figure 2.1.

Figure 2.2
AC Time at Home After AC Transformation

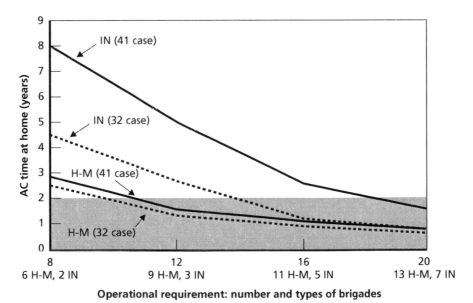

Under transformation, the addition to the rotations of 10 transformed brigades improves the situation somewhat, but problems remain for the higher levels of operational requirements. Even after transformation, the heavy-medium units are under two years time at home whenever total requirements rise to 12 brigades or higher. This is the case because, of the 10 new transformed brigades, only two will be heavy-medium units. Infantry units are able to achieve the two-year goal until requirements reach about 20 brigades. Of course, these transformation results represent some period in the future, if transformation works out as expected. As we argued earlier, in the near term the Army faces a situation somewhere between the baseline and transformation lines—which places considerable stress on the heavy and medium forces if the deployment level and types of units remain where they were in 2004.

Flexibility in Requirements by Unit Type. Because many of the problems we have cited are concentrated within the heavy-medium forces, the Army could gain some advantage if it could be more flexible in stating or meeting requirements for those forces. For example, the situation might evolve in such a way that the combatant commander would be willing to accept more infantry units and fewer heavy-medium units—despite the resulting compromises in armored protection and ground mobility.

In the above cases, we have been assuming that the number and types of brigades in each of the four cases of operational requirements are specified and weighted toward heavy-medium forces. What if the operational requirements were more flexible? By "flexibility," we mean that at least some parts of the stated requirement for one type of unit could be filled by another type. This would be the case if the requirement were fungible or if one were prepared to take the operational risk of assuming infantry units can do what heavy units can.

If the requirements permitted full substitution of unit types, then one could seek to equalize the burden on all the units in the Army's force. The result is shown in Figure 2.3, which represents the case of the transformed force. The line labeled "All" reflects the time at home that would be achieved by all AC units if heavy, medium, and infantry units were treated as interchangeable. This may be contrasted by the lines labeled "IN" and "H-M," which repeat the previous results (namely, the case where infantry and heavy-medium requirements can be met only by the category of unit specified).

By equalizing the burden, the composition of each deployment would reflect the types of units in proportion to the Army force as a whole. For example, in the case of an operational requirement of 16 brigades, rather than meeting the specific demand of 11 heavy-medium and 5 infantry in the previous analysis, the deployment would have included 9 heavy-medium and 7 infantry (the proportional composition of the full AC inventory). Even with this flexibility in the requirements, the Army would face problems in achieving its goal of two years time at home for requirements greater than 14 brigades. For example, at a requirement of 16 brigades met using full flexibility, each unit gets just 1.56 years (about 19 months) at home between deployments.

Figure 2.3
AC Time at Home, Allowing Full Flexibility of Unit Types (After Transformation)

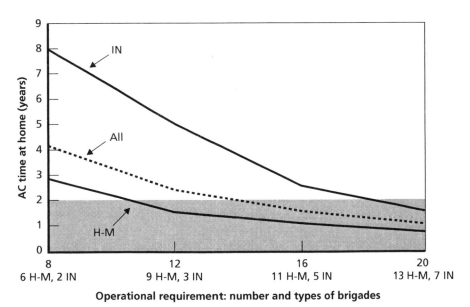

Operational requirement: number and types of brigades

RAND *MG362-2.3*

If, on the other hand, the rotation requirements shifted to emphasize infantry units, the strains on heavy-medium units would decrease. However, infantry units would then be increasingly stressed. For example, consider the case of an overseas rotation requirement of 16 brigades. Assume first that nine are specified to be infantry and seven heavy-medium and that heavy-medium units can only fill the heavy-medium requirement and infantry units only the infantry requirement. The time at home for infantry units would be 12 months and for heavy-medium units 27 months.

Of course, it is possible that heavy-medium units, after filling their own requirement, could fill some of the infantry requirement—a strategy of partial flexibility. In that case, the time at home for infantry units could rise to about 15 months, while for heavy-medium units it would be 24 months. Alternatively, one could allow full flexibility, which, as previously discussed, would yield about 19 months time at home for all unit types. However, what one cannot do in this situation is achieve a time at home of two years for all units. That holds true no matter how the balance of requirements shifts between heavy and lighter forces.

AC Unit Readiness and Availability

The amount of AC time at home affects the readiness of units for additional unplanned contingencies and the ability of Army units to execute their planned training cycles. Here we explain how we assessed the readiness and availability of AC units, and we quantify the results of the overseas rotation cases outlined above in terms of the "number of AC ready brigades."

Assessing AC Unit Readiness. Army units have much more to do than rotating to current overseas operations. Their traditional missions—including fighting and winning the nation's wars—involve a high degree of combat readiness and training across a variety of missions. Every unit has an extensive "mission-essential task list" intended to guide its training activities, and every division and brigade develops a training calendar to manage the cyclical train-up of subordinate echelons (from platoon through battalion, including various task forces and teams combining different types of units).

This training cycle has the goal of producing a brigade that is fully trained and ready to go to war, should circumstances require it. Typically, such a training cycle culminates in a rotation to a Combat Training Center (CTC), in which trainee units conduct simulated operations against a professional opposing force and receive expert feedback from observer-controllers who have specialized in assessing battle outcomes and training.

When operational deployments occur, this peacetime training cycle is apt to be disrupted. Indeed, even training at the CTC has shifted to preparing for deployments to Iraq, and the "opposing force" unit at the National Training Center, the 11th Armored Cavalry Regiment, is deploying to Iraq (Schoomaker, 2004).

To illustrate effects on units, Figure 2.4 exhibits a notional rotation cycle for a heavy unit, including segments during which the unit would attempt to conduct its normal warfighting training while at home.

The full cycle includes a variable-length period at home, followed by a one-year deployment to the theater. The cycle begins after the previous deployment, with a period of recovery that we assume to be two months. During that time the unit may release many of its personnel (for example, relinquishing soldiers who are scheduled to move to another unit, attend Army professional schools, or leave the service). It will also acquire other incoming personnel, and it must refurbish equipment, conduct maintenance and safety checks, and perform myriad tasks to prepare for an intensive train-up cycle. At

Figure 2.4
AC Unit Readiness Cycle

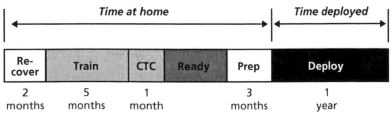

RAND MG362-2.4

this point, the staff will develop detailed plans for the coming training period and for other contingencies that may occur.

After the recovery period, the unit enters a period of training,
first by platoons, then companies and company teams, then battalions and task forces, and so forth. Support units undergo a parallel
process. In Figure 2.4 we allot five months for this training, which is
approximately the amount that was planned in many brigade-training
calendars during the late 1990s and early 2000s. After that, the unit
moves to a CTC (typically, a two-week event with several days before
and after devoted to equipment preparation and movement).

Only after the CTC rotation is accomplished would we regard
the unit as fully "ready." That is, after the training cycle and CTC
rotation, the unit would generally be regarded as capable of deploying
to a combat theater with an appropriate degree of preparation.[8] Figure 2.4 shows this period, but without a fixed length. In effect, the
length of the "ready" period depends on the amount of time at home
afforded by the deployment schedule. The ready period lasts until the
next deployment comes up. We allocate three months for special
preparation for the deployment (e.g., for "mission readiness exercises," outfitting with any special equipment needed, familiarization
with the terrain and culture of the deployment area of operations,
and other mission-specific preparation).

Notice what happens as the amount of time at home is com-
pressed. Consider first a situation in which time at home is two years.
All the other activities in Figure 2.4 occupy 11 months out of that
two-year period. Thus, the unit has a period of 13 months when it is
trained, ready, and available for any other missions. However, if the
length of time at home drops to, say, 11 months, there is *no time*
when the unit is ready but not deployed. Thus, in such a situation the

---

[8] It must be granted, of course, that in a crisis the Army would take a unit that had not
completed its training and deploy it to a wartime theater anyway. The unit would execute as
much training as possible on an accelerated schedule, within the available time constraints.
However, this is only what one would want to accept in an exigency. It is not a prudent basis
for planning the force. We argue that for planning and analysis purposes, we should state the
full amount of training that most commanders and observers would feel is necessary, with
the objectives of maximizing combat power and minimizing casualties.

nation would never have an AC brigade that has fully trained for its wartime mission but not deployed until the deployment requirement receded or other sources of deploying forces could be tapped.

Number of Ready AC Brigades: Baseline Force. With the above plan in mind, it is a simple matter to calculate how many AC brigades can be ready at any point.[9] Basically, the number of ready brigades is proportional to the fraction of total cycle time in which an AC brigade is trained and ready. Figure 2.5 shows the result for the baseline force structure. The types of heavy-medium and infantry brigades are specified in these cases of operational requirements, as we defined earlier.

Consider first the requirement of eight brigades. Under that requirement, both heavy-medium and infantry units enjoy consider-

Figure 2.5
Number of Ready AC Brigades at Home (32 BCTs Rotating)

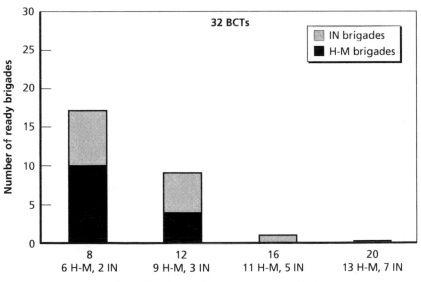

---

[9] See the appendix for the logic and formulas to make the calculation.

able amounts of time at home. The result is that at any given point, the Army has 10 heavy-medium brigades ready and seven infantry brigades ready. The numbers are appreciable, but not large; for example, 10 heavy-medium brigades are equivalent to a little more than three heavy divisions. By comparison, the United States sent more than five divisions to the Persian Gulf War in 1990.

Within Figure 2.5 the most striking result is the rapid decline in available units as requirements rise. At a requirements level of 12, for example, the Army has only four heavy-medium brigades ready (about 1.3 divisions); at higher levels it has none at all. Indeed, at a requirement of 16, the Army has no heavy-medium brigades and only one infantry brigade ready, and at 20 it has no ready brigades of either type.

If, instead, flexibility were permitted in the types of units called for in the overseas rotation, the overall number of ready brigades would not change, but the mix would have more heavy-medium brigades.

We caution again that this does not mean there are no *units* at home station that could be assigned to a new requirement. The problem is that those units may be at home, but they are in the midst of recovery or training. In an emergency, they could (and would) be deployed but clearly at some operational risk.

We have stipulated various overseas rotational requirements and now described the number of units available and ready at home for additional contingencies. It is not possible to posit with certainty in advance what such contingencies might be and therefore what number or types of ready units will be required. Too many uncertainties exist, and many potential demands could arise. The Secretary of Defense has called for military capabilities to be available more rapidly than in the past—i.e., for U.S. forces to deploy to distant threats within 10 days, to defeat an enemy within 30 days, and to recover quickly enough to handle a second fight 30 days later (Sherman, 2004, p. 22). It seems likely that military planning will focus on rogue states seeking to acquire weapons of mass destruction, an emphasis that appears in *The National Security Strategy* (2002, pp. 13–15). Some of these states with nuclear aspirations, such as North

Korea and Iran, possess considerable conventional military capabilities and could present a formidable threat. Planning is also likely to incorporate actions against terrorist networks—e.g., in Pakistan, Indonesia, or Yemen, given that *The National Security Strategy* (2002, pp. 5–6) calls for capabilities to disrupt and destroy terrorist organizations of global reach, preemptively and unilaterally if necessary. Also, DoD guidance is also being developed for using U.S. forces to defend the homeland (Sherman, 2004, pp. 22, 24). As of 2004, the Army had two brigades committed to respond to potential actions involving military assistance to civil disturbances in accordance with the DoD Civil Disturbance Plan. It also had five battalions in the active Army standing ready as rapid and quick-reaction forces to respond to potential domestic homeland security emergencies, such as critical infrastructure protection, counterterrorism operations, and managing consequences of terrorist attacks (Davis et al., 2004, p. 7).

These far-reaching defense goals could call for a wide range of forces. In comparison, the number of ready brigades in Figure 2.5 seems modest. We cannot judge conclusively whether this available number of brigades is sufficient for all the other missions that the nation might assign to the Army, but it could be used as a starting point to assess readiness for any particular set of potential requirements. We are not suggesting that there is any "right" number of ready brigades but rather that many potential demands for ready units exist in the future. The nation as a whole will need to decide in terms of its planning what capabilities it wishes the Army to have for these other contingencies and what risk it is prudent to assume with respect to having the Army able to provide ready capabilities.

Number of Ready AC Brigades: Transformed Force. Perhaps, one may surmise, this troubling situation will be improved after the Army's transformation. Figure 2.6 shows the results for the transformed force structure, when the types of heavy-medium and infantry brigades are specified in the four cases of operational requirements. Transformation does yield several improvements and makes the situation less bleak at the lower requirements. However, even after transformation the Army has only two ready heavy-medium brigades

Figure 2.6
Number of Ready AC Brigades at Home (41 BCT UAs Rotating)

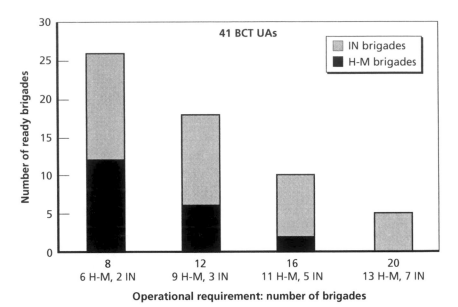

RAND *MG362-2.6*

when 16 brigades are required for rotation and it has no heavy-medium ready brigades when 20 brigades are required.[10]

One way to mitigate the risks would be to shorten the amount of training time or the other "overhead" segments in Figure 2.4 (recovery and deployment preparation). No doubt many such attempts are under way, and with experience the Army may be able to disseminate lessons learned, standardize procedures, and achieve time efficiencies when preparing units that have already undergone similar deployments. However, we argue that such improvements will take time and could prove illusory if the situation changes—for instance, if the nature of the deployed mission gravitates toward a lesser (or greater) intensity of conflict or the locale of conflict shifts to another

---

[10] Although time at home is a little higher in the transformed force than in the baseline force at the requirements level of 20, it still does not reach the level of 11 months and therefore does not produce any ready heavy-medium transformed brigades.

place or cultural context. If recent events have taught anything, it is that the tactics of terrorist organizations can easily shift. Therefore, it is plausible that the United States may in fact face the situation portrayed in these figures for some time. That makes it all the more important to consider how to use the reserve forces, as we will describe in the next chapter.

Life-Cycle Manning Implications

The amount of AC time at home also affects the ability of the Army to implement its evolving program for life-cycle manning of units. Under the life-cycle concept, each brigade[11] is brought up to strength with a complement of soldiers intended to remain with the unit for 36 months. During the first few months of the life cycle, a unit is "reset" with its new soldiers and conducts collective training. On completion of training, the unit is considered ready and subject to deployment at any time until the end of its 36-month life cycle. At the end of a cycle the unit relinquishes first-term and career soldiers who leave the active Army or are reassigned. The unit stands down, and a new 36-month life cycle begins, with the remaining soldiers forming the nucleus around which newly assigned soldiers fill in for the new cycle.

We have conducted some rough calculations to explore the compatibility between life-cycle manning and varying levels of time at home. Because each cycle lasts exactly three years, the plan works well if units can spend two years at home for every year deployed. With two years at home, the unit is able to reset, train, stand ready, and then deploy (for 12 months) once within its 36-month cycle. The life-cycle plan can also fit readily with rotations if units deploy for 18 months and are at home 18 months in a 36-month cycle.[12]

---

[11] Life-cycle manning will affect all the brigades, and an implementation schedule has been established based on current and future operational deployment and redeployment times as well as the modular transformation schedule. See Department of the Army (2005a).

[12] Army information sources describe how, under three-year unit life cycles, a brigade could train, deploy for a shorter period (say nine months), return for a period of several months, then deploy again for a second nine-month deployment before reaching the end of its life cycle. Alternatively, a unit could deploy for a single 18-month period within a 36-month life

However, when deployment tours are 12 months and time at home declines below two years, two serious conflicts arise with the life-cycle schedule. First, each successive rotation moves the deployment date earlier within a unit's life cycle. Eventually this means that the unit faces a training-time deficit: it does not have enough time to fully reset and train before its next scheduled deployment. Second, after one or more cycles the unit eventually faces an infeasible deployment: the unit is scheduled to deploy a second time near the end of the life cycle, even though the cycle has less than one year remaining. In that situation, the Army could deploy the unit anyway and thereby postpone the resetting of the unit, or the unit could deploy for less than one year and thereby risk disrupting the entire deployment schedule for all subsequent units.

These conflicts would take some time to develop. Critical is how early in its life cycle a unit must make its first deployment, and this depends on the length of time the unit can spend at home. Generally, to sustain the life-cycle program it would be desirable to keep units at home for 18 months or longer.[13] This further reinforces the argument that the Army should aim for two years time at home as a key goal for its rotational deployments.

---

cycle. Either strategy would permit the Army to meet demands large enough to require brigades to be deployed 18 out of every 36 months without conflicting with life cycles. Life-cycle manning, in this view, should return to each brigade an experienced cadre of noncommissioned officers and junior personnel with each life cycle, perhaps yielding a reduction in training time to nine months. Under this assumption, and with tight scheduling, a life cycle could consist of four equal nine-month segments: training, first deployment, home, and second deployment.

[13] Assumptions about training time and length of the deployment tour play an important role in determining when these conflicts arise. We assumed 11 months of training time and a 12-month deployment tour. Under these assumptions, if time at home is one year, the life-cycle manning system can be sustained for three years before the appearance of training-time deficits and infeasible deployments. If time at home rises to 15 months, the system can be sustained for three years until units face a training-time deficit and five years until infeasible deployments. If time at home can be lengthened to 18 months, sustainable time is six years until a training-time deficit and nine years until infeasible deployments. And if time at home is 21 months, under our assumptions sustainable time is 12 years until a training-time deficit and 18 years until infeasible deployments. Recent Army planning assumptions allocate only · six months for reset and training. Under those assumptions, these conflicts would not emerge as soon.

CHAPTER THREE

Employing Active and Reserve Components Together

In the analysis of AC unit readiness and life-cycle manning, we have been representing rotational strategies that use only AC brigades. Evidently, adding RC brigades to the rotational mix would lessen stress on the AC. However, there are limitations on how much and how frequently RC forces can be used. This chapter describes the available RC structure, the extent to which RC units can participate in rotations over the long term, and effects of various changes in RC employment policies. It will then briefly describe the effects of sustained overseas rotations on individual soldiers' time at home.

Reserve Force Structure

Table 3.1 summarizes the baseline inventory of RC combat brigades, all of which are located in the National Guard. The existing National Guard force structure has 25 heavy brigades and 12 infantry brigades. Of these, 17 are organized as separate brigades (with 15 organized as enhanced separate brigades or E-brigades). Others are part of National Guard divisions. At present no units are organized as medium (Stryker) brigades.

The distinction between enhanced and divisional brigades is important for two reasons. First, E-brigades have a more comprehensive set of assets organic to their structure, owing to their status as "separate." Therefore, their major subordinate echelons are more apt

Table 3.1
RC Force Structure: Baseline (2004)

Unit Type	RC Force Inventory (Number of Brigades)[a]		
	Separate Brigades	Divisional Brigades	Total
Heavy	8[b]	17	25
SBCT	0	0	0
Infantry	9[b]	3	12
Total	17	20	37

[a]All RC brigades can participate in the rotation.
[b]The eight heavy brigades are E-brigades; seven of the infantry brigades are E-brigades.

to train together as a fully deployable unit and their leadership is more familiar with the range of different elements needed for operational missions. Second, E-brigades have enjoyed priority for resources and training since the early 1990s, under planning assumptions that they would be the first RC combat assets to deploy in an emergency. As a result, the 15 E-brigades are generally regarded as having a higher state of readiness than divisional brigades. Confirmation of that view can be seen from the fact that all but one of the National Guard combat brigades that have been deployed to Afghanistan and Iraq have been E-brigades. It is generally accepted that to deploy the divisional brigades would take more time and resources than the E-brigades.

Transformation of the RC, however, will change this picture in several important ways. Table 3.2 shows the Army's plan for transformation of the National Guard brigade structure, which is to take place from 2005 through 2010 (Department of the Army, 2005a). Two changes stand out. First, the Army plans to eliminate the distinction between E-brigades and divisional brigades and to maintain all RC brigades at comparable readiness levels and equal in capabilities to the AC brigades (Cody, 2004).

Second, after transformation is complete, the plan calls for a much lighter inventory: the National Guard will contain only 10 heavy brigades but will have one Stryker brigade and 23 infantry bri-

Table 3.2
RC Force Structure: After Transformation

Unit Type	RC Force Inventory (Number of Brigades)[a]
Heavy	10
SBCT	1
Infantry	23
Total	34

[a]All RC brigades can participate in the rotation.

gades. Thus the total number will shrink slightly, and the number of brigades designated as heavy will decline dramatically. This reflects the Army's assessment that long-term future requirements for RC forces will gravitate toward missions that can be served by light forces (Department of the Army, 2004b).

Reserve Force Mobilization and Deployment

The Mobilization and Deployment Cycle

A distinctive feature of using RC units is that they must undergo a transition from peacetime status to active duty—a period of mobilization and preparation for the coming mission. When they return home, they likewise have a period of demobilization before returning to their civilian homes and jobs. These transitions complicate the utilization of RC assets, and they need to be taken into account in analyzing how the RC can be used.

At this writing, more than ten brigades have been mobilized for deployment in Afghanistan and Iraq. Therefore we have a modest base of recent experience on which to build a picture of the mobilization and deployment process for RC brigades. Figure 3.1 exhibits our synthesis of a typical mobilization cycle.

When an RC unit is called to active duty, it first must undergo a formal mobilization process in which individuals assemble at a central point, personnel are processed and legally shifted to active-duty status, and the unit conducts "preparation for onward movement."

Figure 3.1
Typical RC Mobilization Cycle

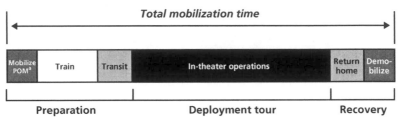

ᵃ POM = preparation for onward movement.
RAND *MG362-3.1*

In most cases—certainly in the case of a large formation, such as a brigade—the unit also undergoes a training process. Training can be lengthy, particularly for RC units, which may not have been together in a field environment since their last annual training period.[1] They must then move to the theater. In this process, more than one move may occur—for example, from home station to a maneuver training area, from there to a railhead or airport where equipment is shipped, and from U.S. locations to the destination overseas.

Only after these tasks have been accomplished does the unit commence its operational mission, as shown by the segment labeled "in-theater operations" in Figure 3.1. Then, after the unit is withdrawn from the theater, it must undergo a further demobilization and transit process to return its people and equipment to home station (or to a maintenance site) and transition the personnel from active duty to inactive status.

Duration of Mobilization and Preparation

The above facts indicate that RC units require a certain amount of "overhead" time for each mobilization. In the first deployments to Afghanistan and Iraq, most units required about six months for the preparatory phase, from the time they were formally mobilized until

---

[1] In addition, many small units within the brigade are apt to have experienced personnel turnover, and their last annual training may have occurred with different people or with some members missing because of individual training or other requirements.

they arrived in the overseas theater.[2] If we allot at least two weeks for postdeployment transit and demobilization, a figure of six months seems a reasonable estimate of the amount of time for an RC brigade to be mobilized but not available for in-theater operations. This places limits on the length of overseas tours for RC troops. At the outset of operations in Afghanistan and Iraq, the Army scheduled RC units for six-month tours in theater. Later, many reserve tours were lengthened to one year as the full extent of the operational requirements were appreciated (GAO, 2004, pp. 17–22). However, after adding about six months for preparation and recovery, those reservists faced a total mobilization period of 18 months or more at one time. Such a lengthy period is burdensome to many, and the length of mobilization periods already has led to concern and proposals for reducing "time away from home" for RC members who are called to duty. Many view even a one-year mobilization period as too long, which is why previous deployments (such as to the Balkans) were limited to six months.

In our judgment, mobilizations for as long as 18 months may be feasible in unusual or emergency circumstances, but it is not wise to plan for long-term utilization of reservists for that long, especially if the same units are on schedule to be recalled at a later date.[3] Therefore, in our analysis we have set the standard mobilization period at one year, recognizing that during that one-year period only six

---

[2] For example, the 81st Armor Brigade was mobilized in November 2003 and deployed into its operational area in April 2004. For all RC brigades mobilized through the end of 2004, the period from mobilization to deployment has ranged from four to seven months, yielding an average of about 5.8 months (Army National Guard, 2005; Associated Press wire service reports).

[3] As a matter of law, the President has various options for mobilizing reserve forces, including a presidential reserve call-up (which is limited to 270 days) and a so-called partial mobilization authority (which is limited to 24 consecutive months and requires a declaration of national emergency). The partial mobilization authority was used after September 2001, so technically the Army can mobilize reservists for two years. However, DoD's reluctance to mobilize reservists for long periods is shown by the fact that the initial mobilizations after September 11 were intended to last just one year, including overseas tours lasting just six months. Further, the implementing guidance for those call-ups limited mobilization orders to 12 months, although it allowed the service secretaries to extend that period to a total of 24 months.

months will be available for service in theater. Later in the analysis we will vary these parameters to determine what happens if we relax these assumptions.

Frequency of Mobilization

Utilization of RC forces is also subject to another limitation: frequency of mobilization. Because reserve service is inherently intended to be a part-time commitment, no one expects that RC forces will be called to duty for back-to-back rotations or even closely spaced rotations. Nonetheless, the exigencies of scheduling have resulted in cases where the same RC units (particularly in high-demand specialties, such as civil affairs and military police, where the inventory is small) have been called on two or more occasions within a few years. For that reason, DoD has issued guidelines that direct the services to plan for using the National Guard on a schedule that results in only *one year of mobilization out of every six*.[4] While this is described as a "planning metric" and not a fixed, inviolable rule, it is clear that DoD policy is firmly established in this direction.[5]

In our analysis, we have implemented this rule in the following way. We specify the fraction of total time, over the long term, in which an RC brigade may be mobilized. We will begin by setting this fraction to one-sixth, although at various points we will experiment with parameters that would lead to more frequent utilization, such as one-fifth or even one-third.

Base Case Parameters for RC Utilization

Together, the above restrictions dictate a set of policy parameters that we will use as our base case for RC utilization. Those parameters are

---

[4] See *President Bush's FY 2006 Defense Budget*: "limit involuntary mobilization of Reserve and Guard individuals to reasonable and sustainable rates, ideally no more than one year of mobilized duty in every six years." See also a memorandum by the Secretary of Defense, July 2003, followed by a report (DoD, 2004a).

[5] Department of the Army (2005c), however, sets the goal in terms of "deployments": "one year deployed and five years at home station" for the National Guard and "one year deployed and four years at home station" for the Army Reserve. This would require a frequency of RC "mobilization" of more than once in six years.

- AC deployment tour (time in theater): 12 months,
- RC deployment tour: six months,
- RC mobilization duration: no more than one year at any one time,
- RC preparation and recovery time: six months, and
- RC mobilization time over the long term: constrained to one year in six years.

The analysis below begins with these parameters and then varies them to illuminate the effects of using the RC on the time of AC units at home. To illustrate these effects, we will use the overseas rotation requirement of 16 brigades, where the mix is initially specified to call for 11 heavy-medium units. We will then permit flexibility in the types of units required. We will use the AC posttransformation force structure (i.e., 41 rotating transformed brigades, labeled "UAs" in the figures). We will also maintain the distinction between heavy-medium units and infantry units and focus on the effects on time at home for AC heavy-medium units, where the greatest stress lies. In the rotating inventory, the Army has 23 heavy-medium transformed brigades in the AC and 11 heavy-medium transformed brigades in the RC.

Analysis Procedure

The above parameters can be used to determine the degree of participation that RC units can attain, given various policy constraints. Here, we illustrate how we made our calculations about RC participation. More details and formulas can be found in the appendix.

For illustration of the method, the first step is to calculate the number of "slots" in the heavy-medium brigade requirement that the RC can fill. In this case, that number is a little less than one slot.[6] The reason is that each RC brigade can be called up only one-sixth of the time over the long term (a frequency of one out of six, according to DoD policy), and furthermore only one-half of that mobilized

---

[6] The actual calculation is: (11 RC brigades available)*(1/6 frequency)*(1/2 time available when mobilized). The result is 0.92 slots filled by the RC at any given point.

time (six months out of a 12-month mobilization) is devoted to filling a slot in theater. Given that the RC can fill about one slot, the AC must fill the remaining 10 slots. Using the logic developed in Chapter Two for AC units, the time at home for AC brigades will be about 1.3 years, or a little over 15 months.[7]

In what follows, we will show results of varying all the parameters just described. We begin with the simplest variation, looking at changes in the number of RC brigades that participate in the rotation schedule.

Varying RC Utilization Policies

Adding RC Units to an AC Rotation Schedule

Figure 3.2 shows the AC time at home that results when RC units are added to the rotation. On the horizontal axis, the figure shows the number of RC transformed heavy-medium brigades (labeled "H-M" in the figure) that are rotating. The leftmost point, labeled zero, indicates the situation when using no RC brigades at all. The corresponding AC time at home, represented by the dot at the bottom left, is 1.09 years or about 13 months—the same result that appeared in Chapter Two, Figure 2.2. (Here we show the Y-axis in months as well as years, to facilitate observing the small differences that appear.)

The line shows how AC time at home increases as RC units are added to the rotation. The middle circle, labeled "Transformed RC H-M brigades," represents the situation when all 11 posttransformation heavy-medium RC brigades are included. As can be seen, this makes a difference, but not a large one. By using all 11 heavy-medium brigades, AC time at home rises from about 13 months to about 15.5 months. If some RC brigades are held out of the program, the yield is even smaller.

---

[7] This logic can be generalized to handle various tour lengths, mobilization frequencies, mobilization durations, and so forth. In the analysis that follows, those parameters will be varied. The appendix shows the details of the methodology.

Figure 3.2
Effects of Adding RC Units to Rotation

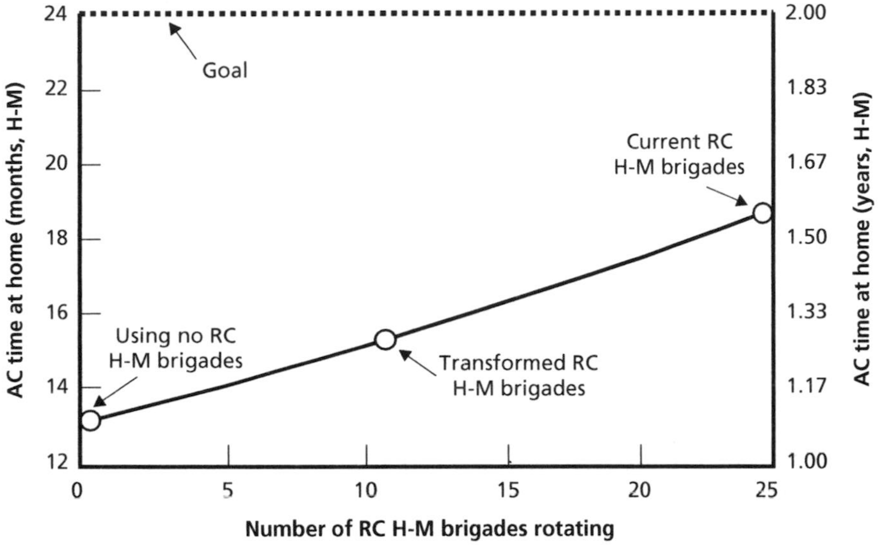

NOTE: Assumptions: 41 UAs rotating; requirement of 16 brigades (11 H-M).
RAND MG362-3.2

By comparison, the situation could improve further if the Army had a larger number of heavy-medium brigades available in the RC. Recall that, at present, the National Guard actually does have more. In fact, the Guard has a total of 25 heavy brigades, counting both separate brigades and those organized within National Guard divisions. Therefore, we extended the line to cover that situation— essentially, what would happen if the Army did not change the National Guard mix of heavy versus light units. If 25 heavy-medium brigades were available in the Guard to meet the 11 heavy-medium requirement, they could absorb more of the deployment requirement and thereby allow AC time at home to lengthen to 19 months.

This is a substantial improvement, but it would require considerable investment. It implies that the Army would keep the current number of National Guard heavy brigades *and* make them equal in capabilities and readiness to those planned for the RC transformed force. Furthermore, note that even under these circumstances AC time at home does not come close to meeting the goal of 24 months (two years at home between deployments).[8]

One other point is worth emphasis here. In all of the above cases, the RC can support only a modest fraction of the requirement in theater. As we have seen, if 11 RC brigades are rotating, they fill about one slot in each rotation whereas the AC fills 10 slots. Thus, the ratio of AC to RC forces in theater is about 10 to 1. This is not an outgrowth of any preference for AC over RC. It simply reflects the practical and policy restrictions on using RC units and personnel too often or for overly long periods.

Changing Reserve Utilization Policies

So far, the options considered have not achieved the two-year goal for time at home of heavy-medium units. What else can be done? One approach would be to try to utilize RC units more efficiently—to gain more yield from RC participation. Here we consider two possible policy changes:

- Reducing RC mobilization and preparation time.
- Mobilizing RC units more frequently.

Mobilization and Preparation Time. The six-month period of preparation before deployments exacts a steep "overhead" cost (in time) from RC mobilizations. What if that period could be shortened? For example, suppose that RC units received more intensive premobilization training for rotational missions or that more extensive resources (e.g., trainers and training facilities) were available to speed up the mobilization process. Such initiatives would be expensive and uncertain,[9] but if they succeeded, how much would they pay off?

---

[8] Utilizing National Guard infantry units in overseas rotations will produce similar effects on the time at home of AC infantry units. In this particular case, though, AC infantry units would have more than two years time at home in all the variations.

[9] For an assessment of how resource-intensive it can be to enhance RC premobilization and postmobilization training, see Sortor et al. (1994) and Lippiatt et al. (1996).

Figure 3.3 shows the effects of shortening RC preparation and recovery time, for the same case: an overall requirement of 16 brigades in which 11 are specified to be heavy-medium. At the left side of the graph lies the point that represents the situation we have just been analyzing: RC brigades require six months for preparation and recovery, leaving the remaining six months of their one-year mobilization period to serve in the overseas theater. Other points on the line show what would happen if preparation and recovery time were reduced and the overseas deployment period were lengthened to fill the rest of the one-year period. For example, the rightmost point on the line represents a situation where preparation and recovery are accomplished in only three months; in that case, the brigade would spend nine months in theater.

Unfortunately, the yields from such an initiative are disappointingly small. Even if preparation and recovery could be shortened to three months instead of six, AC time at home would rise by less

Figure 3.3
Effects of Reducing RC Preparation and Recovery Times

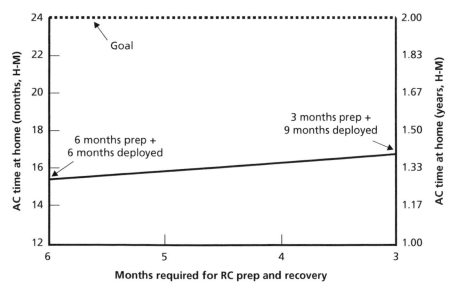

NOTE: Assumptions: 41 UAs rotating; requirement of 16 brigades (11 H-M).
RAND MG362-3.3

than one month. And cutting preparation time that much would be challenging. More feasible changes, such as shortening preparation time by one month, would yield very small gains indeed.

Frequency of Mobilizing RC Units. Suppose instead that DoD were to accept the political and social cost of mobilizing RC units more frequently. Instead of placing the limit on utilizing the National Guard one in every six years, suppose it were one in every five years, four years, or even three years. How much would that buy, in terms of AC time at home?

The answer again is "not much." In Figure 3.4, for the same case: an overall requirement of 16 brigades, of which 11 are heavy-medium, the lower line shows what would happen to AC time at home if RC units still required six months for preparation and recovery, but they were utilized more frequently. For example, using RC units at a rate of one in five years adds only about one-half month to

Figure 3.4
Effects of Using the RC More Frequently

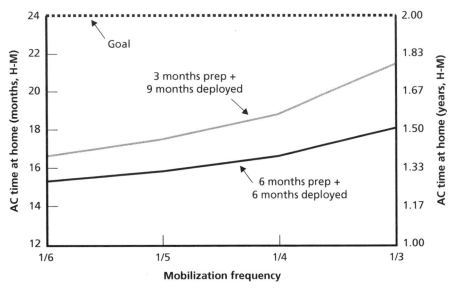

NOTE: Assumptions: 41 UAs rotating; requirement of 16 brigades (11 H-M).
RAND *MG362-3.4*

AC time at home. Even going to a rate of one in four would raise AC time at home by only one month.

The yield would be somewhat better if higher frequency were combined with initiatives to shorten preparation and recovery time. The upper line in Figure 3.4 shows that result. It is obvious that even with the most optimistic assumptions the goal will not be reached. In fact, in our judgment a reasonably optimistic case would be that preparation and recovery could be shortened by one or two months and frequency could be one in five years. Under those circumstances, AC time at home would be a little more than 16 months—less than a month higher than the base case. To achieve AC time at home of two years for heavy-medium units, with our initial assumptions about RC utilization, the rotation requirement would need to drop to 8 heavy-medium brigades. The overall rotation requirement would then depend on the number of infantry units that were needed.

Therefore, we conclude that changes in RC utilization polices can contribute very little on their own.

Permitting Flexibility in Unit Types. In the above cases, we have been assuming that the number and types of brigades in the 16-brigade operational requirement are specified and weighted toward heavy-medium forces. What if the operational requirements were more flexible and therefore all 34 brigades in the transformed National Guard force could contribute to meeting the 16-brigade requirement?

To begin, two assumptions would be required: any type of RC unit would be able to accomplish the mission and each of the National Guard brigades would be equal in capability to those in the AC. In such a case, if all 34 transformed National Guard brigades were mobilized one year in every six years and deployed overseas for six months, then the time at home for the active units would increase to above two years, for an operational requirement of 16 brigades. If the requirement rose to 20 brigades, however, AC time at home would be under 17 months, even with full flexibility. Alternatively, to meet a 20-brigade requirement, some Army planners suggest accepting full flexibility *and* deploying RC brigades for a one-year tour every six years. That would achieve the goal of two years time at home for AC units, but it would require the Army to mobilize the RC brigades for 18 months in every six-year period (equivalent to a mobilization frequency of one in four years).

By permitting flexibility, the burden of the deployment is shared across all the Army's AC units: heavy, medium, and infantry. However, this may not be an operationally attractive solution because it comes with two serious drawbacks. First, as we have noted before, it would carry the operational risk that circumstances in a theater (or a particular location) might call for one type of unit—perhaps involving armored protection, firepower, or mobility—but that type would not be available. Second, the new, transformed National Guard units are not actually in the inventory and will not be for some time. According to Army plans, the full complement of 34 transformed RC brigades will not be available until late in 2010. In the meantime, to field 34 RC brigades, the Army would have to use all of the divisional National Guard brigades, even though many of them are not equipped like AC units and many have not yet begun to make the transition to the transformed brigade structure. Because of these difficulties, it seems unlikely that the Army or the combatant commander would be comfortable with allowing full flexibility involving both the AC and National Guard, at least for some time to come.

Increasing the Supply of AC and RC Units

Substantial changes in managing and using the RC will not accomplish the goal of having an AC time at home of two years, unless one is prepared to accept that future operational requirements can be met by permitting flexibility in the types of units. What then can be done? The Army could attempt to increase the available supply of units. Here we consider the potential effects of increases in AC or RC force structure by adding more heavy-medium brigades. To illustrate the effects, we will use the same overseas rotation requirement of 16 brigades, where the mix is specified to call for 11 heavy-medium brigades.

Figure 3.5 exhibits the effects of changing the number of AC and RC brigades simultaneously. The lowest line, labeled "32 BCTs," represents the case where AC structure is maintained at pretransformation levels while the number of RC brigades varies. Here we show the range of RC brigades from a low of eight (the number of existing E-brigades) to a high of 25 (the total number of heavy RC brigades, including both E-brigades and divisional brigades).

The second line from the bottom, labeled "41 BCT UAs," represents the situation after AC transformation, with 41 AC transformed brigades available to rotate.

As we have seen before, neither of these cases reaches the goal of two years AC time at home. That is, neither rises above the "shaded zone" on the graph. Therefore, we asked, "what if we added *four additional* heavy-medium active transformed brigades"—beyond the planned number of transformed brigades in the posttransformation plan? This could be accomplished by actually "buying" four new

Figure 3.5
Effects of Increasing AC and RC Supply

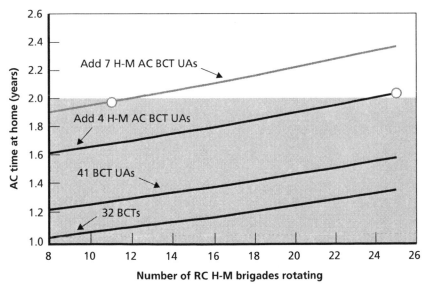

NOTE: Assumptions: 41 UAs rotating; requirement of 16 brigades (11 H-M).
RAND *MG362-3.5*

transformed heavy-medium brigades, or by converting transformed infantry brigades to medium or heavy. Either course would be expensive. Preliminary Army estimates of the initial cost of equipping a new transformed heavy brigade suggest a minimum figure of nearly $1 billion per brigade. Beyond that, further investments would have to be made in basing infrastructure to house the new units and in such training facilities as firing ranges and maneuver areas. Initial infrastructure costs for a transformed heavy brigade could easily exceed $500 million. Thereafter, the Army would incur annual operations costs and personnel costs for about 4,000 soldiers per brigade (unless there were offsetting reductions in infantry units and their personnel were retrained for heavy functions). The annual operating and support costs alone for new transformed heavy brigades are estimated by the Army to be more than $350 million.[10]

What would this buy the Army in terms of AC readiness? It depends on the number of RC brigades available to rotate. If only the eight separate heavy RC brigades are used, AC time at home rises to 1.6 years, still far short of the two-year goal. However, if the Army were to keep and transform its existing 25 heavy RC brigades—rather than convert many of them into infantry, as planned—rotating all 25 would bring AC time at home up to two years.

Of course, such a decision too would be expensive and might conflict with other Army expectations or needs. If transformation of the National Guard continues as planned, the Army would have only 11 transformed heavy-medium brigades in the RC. How many AC transformed brigades would we need to buy then to reach a goal of two years time at home?

The answer is shown by the upper line, in which we represent an Army AC structure that contains *seven more* transformed heavy-medium brigades than planned. (Again, these could be "bought" and added outright to the structure or converted from infantry units.) In

---

[10] RAND estimates of cost are approximate and derived from the Army's FORCES model (March 2005 version) (a limited-access cost-estimating tool) and from CBO (2005, Appendix B, Table B-1). See also CBO (2003), CBO (2004), and Department of the Army (2004b).

this case of an overall 16 brigade requirement, with 11 specified to be heavy-medium brigades, the Army would now have enough brigades rotating to come very close to the goal of two years AC time at home, as shown by the upper line. That is, an AC structure expanded by seven transformed heavy-medium brigades, rotating along with the 11 RC heavy-medium brigades, would allow AC units to spend 1.97 years at home between each deployment.

Thus, we conclude that it is possible to reach the readiness goals for transformed heavy-medium brigades that we have specified. But it cannot be done simply by adding planned RC brigades to the rotation plan, if the Army abides by reasonable utilization constraints. Even measures to improve RC utilization and efficiency will not achieve the goal. Expanding the supply of brigades, however, would make it possible to allow AC units two years time at home and thereby preserve a reasonable number of ready AC brigades available for other missions.

Changing Overseas Tour Lengths

Obviously the length of overseas tours—the amount of time a unit spends in theater—is an important determinant of the above result. The policy debate runs in both directions. Many observers, including those in Congress and the military services, have advocated *shortening* the overseas tour length to ease the burden on soldiers. Others argue, from the combatant commander's point of view, that short tours create too much in-theater turbulence and reduce the experience level of units in theater.[11] Added to this debate is the view of some experienced commanders, who argue that it is best to match the tour lengths for AC and RC units, so that they can train together and operate more smoothly together while overseas.

---

[11] See, for example, an exchange during testimony before the House Armed Services Committee, in which the Army Chief of Staff noted that the Army would prefer to shorten AC tours but could not do so because of operational considerations (U.S. House of Representatives, 2004).

We examined this issue by varying AC and RC tour lengths, to trace potential effects on both components. We found that a continued policy of 12-month tours for the AC and six months for the RC was likely to be the Army's only option that is feasible both politically and militarily, as long as deployment requirements remain high. To explain that conclusion, we consider the result of three cases.

Case A: AC 12-Month Tours, RC Six Months. This reflects our baseline condition: AC units stay in theater for one year, but RC units stay for only six months.

Case B: AC and RC Six-Month Tours. This reflects a policy in which AC tours are deliberately matched to RC tours. Thus, deploying units from the two components could train together and would remain together in theater during the course of a six-month period. A downside, of course, would be more frequent turnover in theater. During much of the time, units would be newly arrived and hampered by lack of familiarity with the surroundings.

Case C: AC and RC 12-Month Tours. This reflects a policy in which, again, AC tours are matched to RC tours but set at one year for both. It gains the benefit of matching AC and RC tour lengths and yields units with more experience in theater but requires longer tours by soldiers and increases RC mobilization time.

How would these changes affect AC time at home and the number of ready AC brigades? Table 3.3 reports outcomes in the situation in which the overseas requirement remains high—that is, an overall requirement of 16 brigades.[12] These results reveal considerable disadvantages of both cases B and C. In case B, AC units must rotate much more frequently to fill the same number of deployment slots as before. The result is to reduce the AC unit's time at home between deployments to less than eight months. That reduces the time for training and as a result prevents any unit from executing a full 11-

---

[12] Conditions for Table 3.3 are those that have been analyzed throughout this chapter: total requirement of 16 brigades (of which 11 are heavy-medium), 23 AC transformed heavy-medium brigades and 11 RC transformed heavy-medium brigades available to rotate, and RC mobilization frequency one year out of every six years.

Table 3.3
Effects of Equalizing Duration of AC and RC Deployments

Case	AC Deployment Duration	RC Deployment Duration	AC Time at Home (months, H-M)	Number of Ready AC Brigades (H-M)
A	12 months	6 months	15.4	4
B	6 months	6 months	7.7	0
C	12 months	12 months	16.2	4

month training cycle (allowing for postdeployment recovery, war-fighting training, and preparation for the next deployment). The result is a catastrophic fall in the number of ready brigades. Because no unit has enough time to train fully, the nation would have no ready heavy-medium brigades available.

Case C avoids that problem, but it adds very little to the AC time at home—raising it from 15 months to 16 months—and it does not increase the number of ready AC brigades. In doing so, it would cause a very different problem for RC units. To remain in theater for 12 months and conduct six months of predeployment training (which is the average time required by RC units up to now), all RC units would need to remain mobilized for 18 months at a time. This is consistent with current deployments to Iraq and Afghanistan but is not a situation that DoD is likely to live with for very long. Furthermore, to sustain the "1/6 mobilization" frequency rule, it would mean calling up units for 18 months for a single mobilization but doing that only once every nine years. That is the only way to sustain the policy of not mobilizing the RC for more than one year out of every six years, on average.[13]

---

[13] It would be possible, in theory, to mobilize RC units for an 18-month period and call them more frequently, say on a six-year cycle. That cycle would mobilize them for 1.5 years, leave them demobilized for the next 4.5 years, and then mobilize them again for another cycle. Another way of describing such a policy would be to "deploy RC units one year in every six years," a formulation found in Department of the Army (2005c). The result would increase AC time at home to 1.5 years—still far short of the goal. Moreover, that would represent an effective frequency of RC mobilization of one in every four years (18 months per six-year period), which conflicts with DoD mobilization policy. In our judgment, such a combination of long mobilization periods, occurring with high frequency, is unlikely to be sustained because RC members, families, and employers would find it too burdensome.

What if demand were to drop? To address that question, we examined the effects of six-month AC and RC deployments in cases where the overseas requirement varied from a low of five brigades up to a level of 16 brigades. The object was to determine how much demand would have to ease to yield two years AC time at home. The results show that total overseas demand would need to be sharply lower: between seven and eight brigades.[14] At those levels the Army would have between nine and 11 AC brigades ready at any given point.

This illustrates how much the Army is boxed in by high levels of operational demand. As we have seen in our earlier analyses, when the overseas rotation requirement rises to high levels (14 or above), many alternative policies become infeasible or carry such profound disadvantages that they become very unattractive.

Effects on Individuals' Time Away from Home

So far, our analysis has concentrated on units' time at home and the readiness of the force as a whole, but what would be the effects on individuals? Here we consider the impact of sustained rotations on individual soldiers—in particular, their amount of "time away from home." The analysis will focus on the total amount of time that a typical soldier spends away from his or her home station—that is, away from family and support facilities that cluster around the unit's home location. There is wide agreement that excessive time away from home undermines soldiers' quality of life and, if continued, could threaten the military's ability to recruit and retain high-quality personnel.[15]

---

[14] If the total requirement is eight brigades (including six heavy-medium), the time at home for AC heavy-medium brigades is 1.76 years and the number of ready heavy-medium brigades is nine. If the total requirement is seven brigades (including five heavy-medium), AC time at home for heavy-medium brigades becomes 2.32 years and the number of AC ready heavy-medium brigades is 11.

[15] As deployments began to rise in the 1990s, official concern about "time away from home" was manifested in several ways. The Army, for example, altered its readiness reporting system

To illustrate the total impact on quality of life, we will estimate the fraction of time that an AC soldier spends away from home *over the course of a military career.* We will focus on soldiers in Military Occupational Specialty 19K, armor crew member. We chose that specialty because it characterizes the experience of personnel in heavy units, which are the most stressed part of the force in our analysis.

Rotational deployments are important contributors to time away from home, but they are not the only events that separate soldiers from their families and home stations. Here we consider the three most important determinants of time away:

- **Deployments overseas.** 19Ks spend much of their time in operational units that will deploy under the rotational plans we considered.
- **Assignments to such permanent stations as Korea.** In the case of Korea, soldiers are sent in unaccompanied status (i.e., without their families). Typically these are one-year tours, shorter than other tours at home or in Europe because of the burden they impose. If past practice is our guide, a 19K soldier during a typical career is likely to have one or more assignments to such locations.
- **Field training.** Units at times conduct maneuvers, gunnery exercises, and other training that takes place in remote locations. In any rotational plan, units that have returned from a deployment will then spend some of their home-station time in the field, adding further to the individual's time away from home.

Field training events last only weeks rather than months, and many would regard the resulting time away as less stressful than longer-term absences. Nonetheless, this time is counted in the Army's

---

to collect monthly data on units' deployments away from home station, called "deployment tempo" (Department of the Army, 1997). Army personnel data were also scrutinized to measure deployment rates for individuals in high-tempo specialties. Congress also enacted military pay provisions calling for special compensation of $100 per day for military members who were subjected to extended periods of time away from home. See Sortor and Polich (2001) and U.S. Congress (1999).

official deployment tempo reporting and has been recognized in the past in congressional action concerning personnel quality of life. Therefore, we included it here.

Another factor in the equation is the interleaving of individual assignments between operational units (also called Table of Organization and Equipment [TOE] units) and the institutional Army (also called Table of Distribution and Allowances [TDA] units). Soldiers in TOE units often deploy, but in TDA units they do not.[16] In a sense, a TDA assignment provides a respite from the three activities listed above. Among junior soldiers (grades E-4 and lower), nearly 90 percent of 19K authorizations are TOE assignments. As their careers mature, however, soldiers are often assigned to TDA units. Among senior enlisted personnel (grades E-5 and higher), about one-third of 19K authorizations are in TDA units. Therefore we separately analyzed the two segments of a typical career—junior and senior—and then combined the two to obtain a picture of a complete career.

Considering all of the above phenomena, we estimated the total amount of time away that would be experienced by a 19K soldier who served a typical career beyond the first term.[17] Here we illustrate the results for the case of the posttransformation force, focusing on heavy-medium units and examining brigade requirement levels ranging from eight to 20 brigades, with the number of heavy-medium brigades we had earlier specified as the requirement. Figure 3.6 shows the percentage of time that a typical 19K soldier could expect to spend away from home over the long term.[18]

---

[16] This is not strictly true for senior NCOs and officers, who may travel to participate in conferences, planning meetings, task forces, and other staff activities while serving in TDA positions. However, the Army personnel system does not capture time for such "temporary duty" assignments, and we judged that it was likely to represent a negligible fraction of time for most enlisted personnel.

[17] See the appendix for calculation methods. Various alternative methods can be used, employing more detail by grade and more assumptions about assignment sequences, but we found that the results from such calculations agreed with results of our approach within two percentage points.

[18] This calculation assumes that 19Ks spend four years in the junior grades and 12 years in the senior grades (the number of years that a career soldier would remain in the 19K specialty after the first term before being promoted into another specialty). If we assumed that

Figure 3.6
Percentage of Time Away from Home over a 19K Career

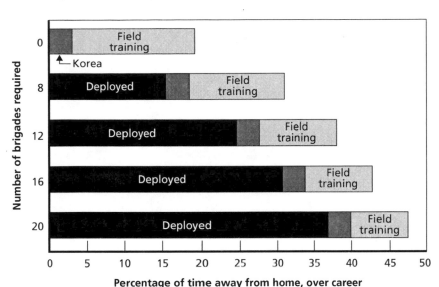

RAND *MG362-3.6*

It is notable that even without deployments, a 19K soldier may expect to spend 19 percent of a career away from home. A small portion of that amount of time is caused by assignments to Korea, but the majority of it stems from field training.[19] As deployment requirements increase, the time taken up by deployments rises sharply—as illustrated by the increasing length of the dark-shaded bars in Figure 3.6. Because the same units are deployed more often, they have less time for field training when they return to their home stations. Therefore, the lightest segments decline somewhat as the requirement rises, but the effects are not enough to offset the rapid increase overall.

___

all soldiers remained for a 20-year career, the bars in this graph would be only slightly shorter (less than one percentage point difference).

[19] Our estimates of field training rates are based on RAND analysis of Army deployment tempo data from the readiness reporting system (Sortor and Polich, 2001). Field training deployment tempo for armor units was reported to be a little less than seven days per month, or 23 percent of the time in field training.

This leads to striking results. At the requirement of 16 brigades—which approximates the current magnitude of overseas operations—the typical 19K soldier could expect to spend 43 percent of his time away from home, over the course of a career. Even at 12 brigades, time away reaches 38 percent.[20] Moreover, most of this time is consumed in one-year, unaccompanied tours, either on deployments or assignments to Korea. This picture contrasts sharply with the career experiences of past cohorts. Until very recently, the typical soldier might have one or perhaps two one-year tours in Korea, and even that was regarded as problematic.

We conclude, therefore, that intense and sustained overseas deployments—requirements of 12 brigades or higher—will place a considerable burden on AC soldiers' quality of life. This burden is far greater than that borne by soldiers during the Cold War or even during the 1990s deployments to the Balkans, which were the subject of considerable debate and concern at the time.

What about RC soldiers? Their situation is simpler to assess because they do not have the same pattern of rotational assignments and extensive field training.[21] A rough calculation of their time away from home might be made as follows. Consider the case where a soldier remains in an RC unit that is mobilized for one year out of every six years, as allowed under DoD planning guidance.

---

[20] These rates of time away from home depend, of course, on the supply of brigades available. This calculation is based on having 23 AC and 11 RC transformed heavy-medium brigades. We made similar calculations for the baseline force (21 AC and 25 RC heavy-medium brigades), and the results turn out to be very similar for the higher requirements levels. In that case, fewer AC but more RC brigades are available. However, if the rotating RC inventory is restricted to the eight heavy enhanced brigades (those providing all but one of the National Guard units mobilized to Iraq and Afghanistan to date) the results become more unfavorable (e.g., time away from home rises to 42 percent for the 12-brigade requirement, 48 percent for the 16-brigade requirement, and 53 percent for the 20-brigade requirement).

[21] Of course, some RC troops come from the AC and others leave the RC to enter on one or more tours in the AC. Here, however, we ignore those complexities and simply examine the experience of a soldier who remains in the RC.

- For five of those six years, the soldier spends approximately 39 days on duty, including 15 days of annual training and one weekend per month of drills. All of the annual training period is away from home. For these high-readiness brigades, we will also assume that during the two years preceding deployment, three of the weekend drills are away from home (e.g., for gunnery or maneuver training).
- During the sixth year, the reservist spends the entire year (365 days) mobilized. This takes the soldier away from home for preparation, training, and overseas deployment.
- The result: A weighted average of 21 percent time away from home.

Therefore, even relatively infrequent mobilizations—one in every six years—impose a considerable burden on the RC soldier. Although civilian jobs may also require substantial travel and periods of absence, few of them involve the danger and stress associated with deployment to a zone of conflict, and few involve a year-long absence. It is evident that many RC troops, when they first entered the service, had no expectation of mobilizations and absences of this magnitude.

As yet, we have no conclusive evidence about the long-term effect of sustained deployments on such key factors as recruiting or reenlistment rates.[22] However, some warning signs on recruiting have appeared. The rate of Army National Guard enlistments, as a fraction of the recruiting goal, has dropped by more than 10 percentage

---

[22] The most comprehensive study (Hosek and Totten, 2002) is based on data from 1993 to 1999. That research revealed that for AC Army personnel in their first enlistment, successive deployments into hostile locations (such as Haiti, Bosnia, and Somalia) generally increased their probability of reenlistment by a small amount. Such was the case, even for personnel experiencing three or more hostile deployments. Reenlistment rates fell, though, when first-termers experiencing three or more hostile deployments also were subjected to two or more nonhostile deployments (such as unaccompanied tours to Korea, disaster relief, and humanitarian aid). For AC Army personnel in their second enlistment, reenlistment rates increased for one or two hostile deployments, but dropped with a third hostile deployment, regardless of the number of nonhostile deployments the soldiers experienced.

points since 2002.[23] Similarly, the AC missed some monthly recruit-ing targets in early 2005.[24] Many observers attribute this in part to the prospect of continuing deployments, but the evidence is not definitive to assess the extent to which the decline may be due to deployments as compared with other factors. Regarding retention, neither the Army National Guard nor the active Army has reported difficulty in meeting reenlistment goals in 2005 (Schultz, 2005; Myers, 2005). However, the retention picture is clouded by the imposition of stop-loss orders and adjustments in retention goals. So although it is too soon to reach definitive conclusions, some signs of trouble are brewing, and it seems plausible that continued high deployment rates could undercut Army manning. In response, to sustain manpower levels, the Army is requesting additional funds to provide larger incentives for retention and recruiting.

---

[23] According to Moniz (2005), the Army National Guard in 2004 fell nearly 7,000 short of its goal to recruit 56,000 soldiers. Their recruiting goal in 2005 was more ambitious, up to 63,000 soldiers. In the year that began in October 2004, they were 24 percent below their target.

[24] According to the Secretary of the Army Francis Harvey (2005), "for the year, we're at 94 percent of where we should be. And, of course, this year's goal in the active is 80,000. And that's up from . . . 77,000. . . . The retention, by the way, is just about on goal."

CHAPTER FOUR

Planning Alternatives

This analysis indicates that the nation faces difficult trade-offs among competing goals in supporting intensive and continued overseas deployments. What are the primary choices to improve the readiness of the Army's active units? In this chapter, we distill the preceding results into a set of leading planning alternatives. All of these involve some type of significant costs or risks.

Long-Term Policy Options

Over the long term, our analysis suggests four general policy alternatives available to the Army for managing intensive deployments and maintaining readiness.

- **Place primary responsibility on the AC.** With this policy, the Army would rely primarily or entirely on the AC to support overseas rotational deployments. It would thus take the risks entailed by having AC units at home for only a short time between deployments and the risk of having few ready AC brigades.
- **Rely heavily on the RC.** In contrast, with this policy the Army could relieve pressure on the AC, enjoy higher levels of readiness, and keep more ready brigades available. However, it would also have to accept the political and resource costs of frequently calling up RC troops and their units.

- **Permit flexibility in types of units for deployments.** If requirements were more flexible and different types of units could be used interchangeably, the Army could equalize the rotation burden and relieve some of the stress on the heavy force. However, such a policy would take the risk of not meeting operational requirements if ground mobility and armored protection are important.
- **Buy new AC force structure or change the mix of AC unit types.** Acquiring more force structure would alleviate the deployment burden and produce additional ready units. However, it would require the nation to dedicate additional resources to acquire and support the new structure.

Table 4.1 defines options and then provides some quantitative information, including outcomes and cost estimates, to help assess the pros and cons of these policy alternatives. It is organized in four panels with white and gray shading, corresponding to the above four policy alternatives. Each of the options is evaluated for a rotation requirement of 16 brigades with the posttransformation Army force structure in place.

We initially weight the requirement toward heavy-medium units and focus on the time at home of these units. Although future requirements are uncertain, the Army has so far emphasized heavy units in its response to the insurgency in Iraq. Among those brigades at home "ready" for other contingencies, heavy units would be critically important in responding to contingencies involving major conflicts, two of which appear in current DoD planning. We do include, though, an option that permits flexibility about the types of units required to fulfill the rotation requirement of 16 brigades.

What emerges in stark relief is the difficulty that the Army faces and the complexity of making decisions that could improve AC time at home and the number of ready brigades.

Option A follows the course of using only the AC. As we have seen before, the result is not satisfactory. Under that plan, AC heavy-medium brigades spend only 13 months at home between each 12-

Table 4.1
Assessing Policy Alternatives

Option	Policy	Resources	RC Utilization	AC Time at Home (Years)	Number of Ready AC Heavy Brigades	Estimated Minimum Startup Costs ($ Millions)	Estimated Minimum Annual Costs ($ Millions)
A	Use AC only; no use of RC	Planned AC only	None	1.09	2	—	—
B	Use planned RC per current policy	11 RC brigades in plan	1/6 years; 6 months prep plus 6 deployed	1.28	4	—	300
C	Use planned RC more heavily than in current policy	11 RC brigades in plan	1/5 years; 3 months prep plus 9 deployed	1.46	5	—	500
D	Keep current RC brigades and use per current policy	25 RC brigades; beef up readiness	1/6 years; 6 months prep plus 6 deployed	1.58	6	3,800	1,000
E	Keep current RC brigades and use heavily	25 RC brigades; beef up readiness	1/5 years; 3 months prep plus 9 deployed	2.17	9	3,800	1,500
F	Permit flexibility in unit types, use planned RC per current policy	34 RC brigades in plan	1/6 years; 6 months prep plus 6 deployed	2.12	9	—	800
G	Add AC force structure or change unit types	Add 7 heavy transformed brigades; 11 RC brigades in plan	1/6 years; 6 months prep plus 6 deployed	1.98	11	5,200–10,000	1,700–2,800

NOTE: Assumptions: 16 overseas brigade requirement, 11 of which are heavy-medium brigades; rotating 41 AC transformed brigades (23 heavy-medium). RAND estimates of cost are approximate and should be viewed as minimums. They are derived from the Army's FORCES model (March 2005 version) and are expressed in FY 2005 dollars.

month rotation to the overseas theater. Furthermore, the nation has only two fully ready AC heavy-medium brigades available at any time—that is, brigades that have undergone a full cycle of training and a CTC rotation. Option A shows no additional costs because it is our baseline for measuring costs of the other options.[1]

The options labeled B through E represent various ways of relying more heavily on the RC, which also entails political and resource costs. The first possibility, Option B, would be to use the transformed RC brigades—including 11 heavy-medium brigades in the plan—under constraints consistent with current RC utilization goals. That is, RC brigades would be mobilized for only one year in every six years, each mobilization would last one year, and that mobilization would include six months of preparation and recovery plus six months of in-theater operations. As columns five and six show, this does not go very far toward our goals; AC heavy-medium units still have only 1.28 years at home (15 months) and the Army gains only two additional ready heavy-medium brigades.

In terms of costs, on average in Option B, almost one National Guard brigade is deployed overseas at any one time.[2] This would incur about $300 million a year in operating costs above the baseline Option A.[3] (The costs in this section should be considered minimums, or lower bounds, for each option because they exclude one or more factors that are difficult to estimate. They do provide, however, a rough guide to the relative costs of the options.)

---

[1] In estimating the costs of the other options, we used the Army's FORCES model (March 2005 version) and expressed costs in FY 2005 dollars. See the appendix for a more detailed discussion of the costs included in each of the options and a summary of additional potential costs that could not be estimated.

[2] Note that our formal analytic approach for examining the policy options involves calculating, in each case, the parameter: the number of slots in the rotational requirement that RC brigades can fill (on a steady-state basis). We describe the logic and algebra behind this calculation in the section of the appendix labeled "Calculation Methods."

[3] See the appendix for how we derive the annual operating costs of mobilizing the brigades in terms of the RC contribution. As described in the appendix, these costs could in actuality be considerably higher because of additional personnel costs not included in the Army FORCES model. For example, retirement and health-care benefits are not included but could easily raise personnel costs by 40 percent or more.

Option C experiments with using the RC more intensively. It retains the RC force structure in the plan but allows RC units to be mobilized for one year in every five years. It also assumes that the Army could make major improvements in preparation and recovery time. It allots only three months for those activities and thereby allows the unit to serve nine months in theater. Unfortunately, although it does "move the needle" somewhat in the positive direction, the results remain meager. AC heavy-medium units have 1.46 years (18 months) time at home, and the Army has five ready AC heavy-medium brigades.

Whether the Army could achieve these results is quite uncertain, but making the attempt would surely impose additional costs. At a minimum, the Army would have to pay for a larger number of National Guard units mobilized, on average about 1.5 brigades each year. Mobilizing these units would cost almost $500 million in annual operating budgets. In addition, cutting the time to train the units after mobilization from six months to three months would require significant changes of two types: more premobilization training on the part of the units and investments in the Army's training infrastructure (facilities, ranges, and instructors) for postmobilization training. It is very difficult to predict the resources necessary for these changes, but they could be significant.[4] Therefore, our estimate for Option C of a little over $500 million is almost certainly low.

To attempt further improvements, Options D and E show what would happen if the Army were to forgo its plan to shift RC units toward a lighter mix. Instead, under these options the Army would retain an inventory of 25 heavy-medium National Guard brigades. It would need to invest in the existing divisional brigades so that they (and the existing separate brigades) would attain the capabilities of future transformed heavy brigades. Even so, Option D shows that the result still falls short if we stick to the constraints of current RC utilization goals. If RC units are called only once every six years and

---

[4] We did include a little more than $25 million to support a year's worth of additional premobilization training for each unit, but postmobilization training was too uncertain to assess. See the appendix for details.

preparation times do not improve, the result is 1.58 years (18 months) time at home for AC heavy-medium units and six ready AC heavy-medium brigades.

This option would have several major effects on cost. First, the Army would be required to convert 14 more RC heavy brigades to transformed heavy brigades than it would under Option A. If those brigades were fully equipped today (which they are not), the Army's costing model implies that the conversion would impose a start-up cost of $3.8 billion. Second, the Army would incur at least $1 billion in additional operating costs each year because heavy brigades consume more resources than infantry units and because more of them would be mobilized each year. Moreover, these estimates are minimums. The actual costs could eventually rise considerably higher.[5]

Option E, in contrast, does achieve the time at home goal (2.17 years at home for heavy AC units). And it provides a total of nine ready AC heavy-medium brigades available for other national purposes at any time. However, this result depends on the minimum start-up investment of about $3.8 billion just mentioned, plus investments to shorten preparation time, plus a willingness to use the RC more frequently than the goals of current DoD policy allows. It is also uncertain because the abbreviated preparation times might never be achieved and frequent RC call-ups might hinder RC recruiting and retention while testing public patience. In this option, the Guard would have an average of almost four brigades mobilized each year, incurring an annual operating cost of about $1.5 billion.

The next row, Option F, assumes something quite different: namely, that heavy, medium, and infantry forces in both the AC and RC can be used interchangeably in the theater. This might be the case, for example, if the units were performing routine security or assistance tasks in a relatively benign environment and if the primary

---

[5] The appendix provides details and also notes factors that might increase the costs substantially, especially the costs to fully staff and equip these brigades. The start-up costs, for example, could easily be higher than $3.8 billion if the divisional RC units needed to replace older-generation equipment with modern equipment or fill shortfalls in equipment if they do not have their full complement today. The upper bound for start-up costs could be $14 billion if the Army had to buy all new equipment for these brigades.

requirement were for large numbers of soldiers (rather than combat mobility, firepower, and protection). Under that situation, the Army could use all 34 RC units in the posttransformation inventory (and the 41 active transformed brigades) to meet the total rotational requirement of 16 brigades. With that supply of forces and no restrictions on unit types, AC units would spend 2.12 years at home between deployments, and the nation would have nine heavy-medium AC units ready at all times. On the downside, however, is the operational risk that the Army might find the wrong type of units in theater (or at a particular hot spot in theater) if tensions rose or the situation deteriorated.[6] Under this option, the Guard would have an average of about three brigades mobilized each year, which would incur additional annual operating costs of more than $800 million.

Finally, Option G represents a decision to expand heavy AC Army force structure while keeping the planned RC structure in place and using the RC in accordance with current RC utilization goals. This could be done by either adding new units or changing the mix of infantry and heavy-medium units in the Army's current transformation plan of 43 brigades. To achieve about two years AC time at home, the Army, according to our calculations, would need an additional seven active transformed heavy brigades. The Army is already considering the possibility of adding five new brigades, but these are currently planned to be infantry units.

It hardly needs to be said that creating new transformed brigades would entail substantial costs, both initial costs (such as personnel, equipment, and base infrastructure) and recurring costs (such as personnel and consumables). If the Army decided to add seven new transformed heavy brigades to its AC force structure by increasing end strength, it would incur start-up costs of at least $10 billion to man, equip, and train the new units and also to build base and training facilities. In addition, operating the seven new transformed heavy brigades would cost an additional $2.5 billion each year. Moreover, these figures represent just the costs of the combat brigades, not other

---

[6] DoD has come under criticism for the deployment of forces to Iraq without adequate armor for their protection. See, e.g., Klein (2004).

support units or institutional structure that the Army might need to go with them.[7] Altogether, the operating costs for this approach would be $2.8 billion because there would be almost one National Guard brigade mobilized each year.

The costs would be lower if infantry units were instead transformed into heavy brigades. The costing model estimates that this would require start-up investments of about $5.2 billion to convert the units. Under this plan, annual operating costs would increase by $1.7 billion because it would cost more to operate the transformed heavy brigades and the nearly one National Guard brigade that would be mobilized each year.

Thus, Option G would carry very substantial costs for the nation. However, the option would meet all rotational requirements, achieve AC readiness goals, provide the types of units we specified in the requirements, and use the RC at reasonable rates consistent with DoD's policy for RC utilization.

In each of these policy options, the transformed infantry units in the AC force would have at least two years time at home. The nation would also have available at least five AC ready transformed infantry brigades and as many as 16.

We do not advocate any one of these options as the best or only solution. Many more possibilities and many intermediate choices would yield some but not all of the benefits desired. What is clear is that no single policy is likely to meet all DoD goals. Each choice involves sacrificing something important or incurring substantial costs.

Near-Term Policy Adaptations

Given the difficulty, costs, and risks of the planning options above, it may be necessary to consider what can be done in the interim to ameliorate the situation and move at least partway toward the goal.

---

[7] See the appendix for discussion of these potential costs, which could be large. For example, historically Army support units have accounted for more soldiers than combat units do; that pattern, if replicated, could double the cost of additional manpower.

So we asked the question: what changes could the Army adopt now, even though their results might fall somewhat short of the goal of two years AC time at home?

Here we consider a near-term base case and three potential adaptations that the Army might institute, without stretching as far as the longer-term changes we discussed above. Each adaptation is considered cumulatively—that is, assuming the preceding ones have already been made.

Base Case. Because we are considering the situation in the near term, we assume a requirement for 16 brigades in theater, of which we specify 11 to be heavy-medium, consistent with Army planning in the spring 2005 deployments to Iraq and Afghanistan. We also assume an inventory that approximates the Army plan for 2007. By that time, the Army expects to have an AC inventory of 43 transformed brigades (including 23 transformed heavy brigades), but the RC transformation will have only just begun. Because the Army has thus far deployed National Guard brigades drawn mostly from the pool of E-brigades, in the base case we assume the same. Thus, the RC rotating inventory would include only the eight heavy E-brigades. Under those circumstances, AC time at home for heavy-medium units would be 14.7 months, and the Army would have three heavy-medium brigades ready at any time.

Employ Some AC Infantry Brigades to Meet Heavy-Medium Requirements. As a first adaptation, some AC brigades classified as infantry might be adapted or reclassified to meet a small portion of the heavy-medium requirement. For example, the 101st Airborne Division has extensive firepower and mobility.[8] With perhaps modest supplementation of vehicles, other equipment, and training, it might be deployed in place of a Stryker or heavy unit. In effect, such a course would represent partial "flexibility" in meeting heavy-medium requirements and would add four brigades to the heavy-medium AC pool. Because it would utilize infantry units that already possess considerable resources and would substitute only about one brigade

---

[8] DoD is planning to deploy the 101st Airborne Division to Iraq in 2006. See DoD (2004c).

within each deployment, it would minimize the operational risk of having an inappropriate mix of unit types in theater. The result would be an appreciable improvement: AC heavy-medium time at home would rise to 19.4 months, and the Army would have seven heavy-medium AC ready brigades at any time.

Expand the Number of RC Brigades Participating in Rotations. To date the Army has deployed only one National Guard brigade not previously classified as an E-brigade, so it still has 16 other heavy brigades in National Guard divisions. Some of the E-brigades are scheduled for transformation in the near term, but conversion of the other heavy units will take some time. In the meantime, if the Army were to invest now in the readiness of three divisional brigades (to make them closer to the capability of E-brigades), the Army could reach the level of 11 available heavy brigades in the RC inventory. If 11 heavy RC brigades were available and the AC infantry brigades were substituted as outlined above, AC time at home for heavy-medium brigades would rise to 20.1 months.

Reduce RC Preparation and Recovery Time by a Modest Amount. By advance planning and more intensive premobilization training, RC brigades might be made more ready for overseas rotations, and thereby use less time in postmobilization preparation. Suppose that preparation time could be cut by a modest amount, say from six months to four months. One RC brigade has already met that timeline. In combination with the above two initiatives, this change would increase AC time at home for heavy-medium brigades to 21.1 months and yield eight heavy-medium AC units ready at any given time.

Together, these measures would require several investments and some recurring costs. For example, the four AC transformed infantry brigades might require supplemental equipment or personnel. The three National Guard divisional brigades would need increased resources to match the E-brigades, and all 11 National Guard brigades in the rotation plan would have to achieve additional readiness to be able to prepare for deployment in four months rather than six months.

On the upside, however, these changes could be made using units that exist in the current inventory, and they are generally consistent with the Army's broader transformation plans. Together they would not boost AC time at home completely to the two-year goal, but they would come reasonably close (21 months). They would also test the capacity to increase RC readiness and contributions to deployments, as well as test the AC's capacity to continue operations under a modest compromise with its two-year goal. These benefits would not be as extensive as the long-term planning options discussed earlier, but they could offer a mechanism for managing the near-term demands of intensive and continuing rotations.

CHAPTER FIVE

Conclusions

The Challenge and Plans to Meet It

The U.S. Army is called on to undertake many types of missions, which are both uncertain and highly variable. Those missions may involve fighting overseas terrorism; defending the U.S. homeland; bringing stability to Iraq, Afghanistan, and possibly other countries; and responding in force to potential conflicts or emergencies in many parts of the world.

Recent events have shown how different and complex these operations can be. For example, we have seen how demanding it is to conduct continuing overseas deployments, as in Iraq and Afghanistan. Meanwhile, the Army must help protect bridges, sensitive sites, and other infrastructure at home. But in addition, the Army must balance these immediate missions against other, longer-term goals. While deploying units abroad, it must still maintain the training and readiness of units at home, which may themselves be needed to deploy quickly for a variety of different threats and emergencies. It must preserve its manpower base through successful recruiting and retention in both the AC and the RC. And it must ensure that future generations of soldiers get proper training for both warfighting and stability operations.

To meet these uncertain and varied requirements, the Army has extensive plans for transformation. For example, the Army is converting its active and reserve forces into modular brigades, increasing the number of AC brigades from 33 to 43, and giving them capabilities equal to those of the current brigades. It is also converting many of the National Guard brigades to infantry brigades and resourcing them to make them comparable to similar AC brigades. The AC plans come with an estimated cost of $48 billion and with an increase in the end strength of the operational Army of 30,000 soldiers. To protect unit readiness and quality of life, the Army aims to have AC brigades away only one year out of three years and to mobilize RC brigades no more than one year in six years. Finally, the Army is also instituting 36-month life-cycle manning in its AC units, with the aim of improving stability and cohesion.

However, notwithstanding these planned responses, we have seen that the pace of recent operations has placed considerable strain on the Army's operational units and soldiers. In an environment of high demand for deployments, AC units must rotate quickly between the United States and overseas theaters; they can conduct only limited training at home; and their soldiers are away from home much of the time. Most of the National Guard separate brigades have been mobilized and their overseas tours are lasting as long as 12 months. Moreover, intense overseas rotations seem likely to continue for the foreseeable future.

Therefore, in our conclusions we consider a series of future conditions that could emerge and pose questions about how the Army can adapt so that it can meet its immediate operational requirements and sustain its force over the longer term. In essence, we ask the questions: given its existing structure and enhancement plans, what missions will the Army be able to execute, and if it cannot meet all needs under current plans, how could it adapt to improve the situation?

Varying Conditions for the Future

Suppose, Initially, That Overseas Rotation Requirements Drop Back to Ten Brigades. With that demand, and assuming that the Army both has the resources to implement its AC and RC transformation plans and can draw on all the RC brigades one year in every six years, all types of AC Army units would have at least two years at home between deployments. The Army would have more than 20 brigades ready for other contingencies, of which at least 11 would be heavy-medium units. The ten-brigade level of requirements is considerably higher than the 1990s average of four brigades but well below the more than 16 brigades deployed in Iraq and Afghanistan in 2005.

The issue for the nation is whether one is comfortable basing future Army planning on this lower level of overseas rotational requirements. This assumption could be plausible if one views the current requirements in Iraq and Afghanistan as an aberration or something to be endured for a short time now or only periodically in the future.

Alternatively: What If High Overseas Rotation Requirements Continue for Some Time? To meet requirements levels in the upper range that we have considered—14 to 20 brigades—the Army would experience serious problems in AC unit readiness and the nation would have few if any ready AC brigades to turn to in a crisis. Transforming the Army into the planned structure of 43 active transformed brigades will help. But transformation is largely in the future, comes with its own uncertainties, and cannot meet the full demand for rotational forces by itself.

The nation could decide to live with these low levels of ready AC units and training time—if it believed that the Army will only rarely need to respond quickly to contingencies with large numbers of forces. For example, one could assume that the Army will need to quickly deploy only small numbers of soldiers with very specialized skills but not large formations with extensive warfighting capabilities. Or one could assume that the Army will have considerable time to deploy large numbers of brigades, as it had during the two wars with Iraq. A parallel assumption would be needed regarding domestic requirements. For example, civilian agencies would be able to assume most of the responsibility for responding to terrorist attacks and other emergencies at home. In effect, this course means assuming that international or domestic contingencies will not require Army combat brigades to do much beyond supporting overseas rotations.

What If the Risks Are Too High for the Army to Plan for Low Levels of Contingency Requirements? As we have described, two adaptations are possible. The Army could turn to the RC and plan on utilizing its units at reasonable rates—e.g., mobilizing all RC brigades for one year out of every six years. However, RC units can be called only at reasonable intervals and can cover only a modest portion of the requirement for overseas forces, even assuming, as the Army is, that all transformed RC brigades will be capable of participating in the rotations. Alternatively, the Army could plan to fill rotational requirements based on the assumption that any unit could fulfill the mission. Such flexibility greatly improves the situation, but only if the transformed National Guard brigades are all available to be mobilized one year in every six years and are all equally capable of meeting the overseas requirements. Such a course carries operational risk, if the theater environment is not benign or missions require armor protection and on-the-ground mobility. To date, the Army has hedged against such risks by deploying forces to Iraq that are predominantly heavy. Moreover, when overseas rotation requirements increase beyond about 17 brigades, AC time at home falls below two years even assuming such flexibility.

What If It Is Too Risky to Assume That Infantry, Medium, and Heavy Units in the AC and RC Can Substitute for One Another in Future Missions? We have explored two options to respond under those circumstances. One avenue is for the Army to forgo its transformation plans to convert heavy National Guard units to infantry units. This would also require the Army to find the resources to make these units—including the divisional brigades—equal in readiness to AC brigades. Alternatively, the Army could take an approach that pursues its National Guard transformation plans and keeps RC utilization within current policy constraints but adds heavy force structure to the AC. This could be accomplished either by changing the mix of the units planned in the Army's transformation or adding additional transformed heavy brigades. But this would call for finding billions of dollars well beyond the current Army modularity plan and would take years to achieve.

The difficulty for the Army is that none of these approaches offers relief today. For the near term, we have suggested some minor adjustments the Army could make. For example, the Army could supplement the capabilities of a small number of infantry brigades and substitute them for heavy or medium units. It could also dig deeper into the National Guard's heavy brigade inventory for overseas rotations than it has to date. By increasing those brigades' readiness, it could relieve some of the burden on the AC and spread the burden more widely across the National Guard. These changes might raise time at home for heavy AC units to 21 months but again not without noticeable costs and some compromises with Army long-term goals.

What is clear is that any approach is fraught with risks and uncertainties. To decide on an overall approach for the future will require the nation to confront a number of trade-offs in terms of the Army's reliance on the AC and RC, on the risks it is willing to take in terms of the Army's ability to meet different types of future contingencies, on what types of training of Army units will be required for different types of operations, and on what resources are available for transforming the RC and increasing AC force structure. Our analysis suggests that the challenge is profound and that making the trade-offs will not be easy.

Unit Types and Calculation Methods

This appendix provides backup documentation on four points: it identifies the brigades assigned to each unit type, using the Army's classification mechanism; it outlines methods used to calculate RC utilization rates, AC time at home, and number of ready AC brigades; it describes methods used to calculate individual soldier time away from home outlined in Chapter Three; and it provides the cost analysis that underlies the discussion of the long-term planning options in Chapter Four.

Unit Types

AC Structure
The existing AC structure, before transformation, is classified as shown in Table A.1.[1] Almost all these units are in transformation or expected to undergo significant changes in the next few years, but we list their sources under the names by which they are widely known.

---

[1] The 11th Armored Cavalry Regiment is not included in our list of deploying units. It is not a fully equipped or manned BCT equivalent to the other heavy units in the force structure but is manned and equipped specifically to serve as the opposing force, a training element, at the National Training Center. The personnel in the 11th Armored Cavalry Regiment, just as those in other uniquely manned and equipped elements (such as the Old Guard), do constitute a well-trained and deployable capability and did deploy to Iraq. The unit, however, would normally be used in its normal training role instead of as a regular element of the deploying force structure.

Table A.1
Classification of Existing Units by Type

Unit Type	Inven-tory	Sources		
Heavy	18	3 brigades each: 1st Armor Division, 1st Cavalry Division, 1st Infantry Division, 3rd Infantry Division, 4th Infantry Division	2 brigades in 2nd Infantry Division	3rd Armored Cavalry Regiment
Medium (Stryker)	4	1 brigade in 25th Infantry Division 1 brigade in 2nd Infantry Division	172d Infantry Brigade	2nd Armored Cavalry Regiment
Infantry	11	3 brigades each: 82nd Airborne Division, 101st Airborne Division	2 brigades each: 10th Mountain Division, 25th Infantry Division	173d Airborne Brigade

We include in the medium (Stryker) category the 2nd Armored Cavalry Regiment because of its wheeled inventory and mission capability, even though in 2003 it had not yet been converted.

Depicting the future transformed structure is more ambiguous because plans vary for future time frames. The analysis in this report relies on the Army's plan for the interim transformed force, which is to be in existence by the end of 2007. It is to contain 43 transformed brigades (20 heavy, 5 medium [Stryker], and 18 infantry transformed brigades).

RC Structure

The RC brigade structure includes two main elements, both resident in the Army National Guard. Until recently these two elements were formally distinguished and resourced separately. The priority element included 17 separate brigades (with 15 E-brigades), which are not part of a National Guard division structure. Also, 20 other brigades are embedded in National Guard divisions. The 15 E-brigades received priority in resourcing and other ways, and they typically contain more capability because of their status as separate brigades.

Table A.2
Existing RC Brigade Inventory

Unit Type	Separate Brigades	Divisional Brigades	Total
Heavy	8[a]	17	25
Medium (Stryker)	0	0	0
Infantry	9[a]	3	12
Total	17	20	37

[a]The eight heavy brigades are E-brigades; seven of the infantry brigades are E-brigades.

Table A.2 shows the number of separate and divisional brigades in the inventory, as of this writing. The Army plans to eliminate the distinction between the two in the future and to transform them into brigades with a very different mix of types, as explained in Chapters One and Two. The result will be ten heavy, one medium (Stryker), and 23 infantry brigades.

Calculation Methods

The analysis in this report is based on computations of the long-run average values for time at home and number of ready brigades. Over a long period of time, these averages should prove valid, although at any given point the short-term experience may vary slightly because integer numbers of available brigades do not fit cleanly into the integer number of units required.[2] Here we describe the algebraic method used to derive the numerical results.

Basic Logic
The method is based on the following logic, including five steps:

---

[2] For example, in scheduling 23 AC heavy-medium brigades to fill a rotation requirement of 11 heavy-medium brigades, the length of each brigade's cycle is 2.09 years. That implies that most brigades will deploy every two years and a small number more frequently, in such a way that the long-run average is 2.09 years.

- **RC utilization.** Specify RC utilization policy: The goal for the long-term fraction of time that RC units are to be deployed overseas, the overseas tour length, and the length of mobilization. For example, for most cases we specified that RC brigades could be deployed at most one out of six years, over the long term; each overseas deployment would last six months; and the mobilization period would last one year (to allow a six-month period of preparation before deployment).
- **RC portion of requirement.** Determine the number of "slots" in the requirement that RC units can fill, given utilization policy. That is, based on the frequency of RC utilization, calculate the number of RC brigades that can be deployed at any given time. That number represents the number of brigade slots in the requirement that the RC can fill.
- **AC portion of requirement.** Subtract the number of slots filled by the RC from the total requirement (the number of brigades required in each rotation). This leaves the remainder as a requirement to be filled by the AC.
- **AC time at home.** Calculate the resulting AC time at home between deployments for a typical brigade.
- **AC ready brigades.** Calculate the number of AC brigades that will be fully ready at any given time.

Parameters and Formulas

The steps in the calculation can be represented algebraically as follows. Parameters pertaining to the RC are shown in lower-case letters, while parameters for the AC are in upper case.

Inputs

a   Number of RC brigades available for rotation

f   Fraction of time each RC brigade may be mobilized, over the long term

d   Duration of an RC deployment (time in overseas theater, in years)

m Duration of an RC mobilization (time, in years, from call-up to completion of active duty)

B Total number of brigades required in a rotation
P Time required by an AC unit for postdeployment recovery, train-up, and preparation for deployment (in years)
A Number of AC brigades available for rotation
D Duration of an AC deployment (in years).

Calculated Values

s Number of slots in the requirement that RC brigades can fill
R Number of brigades in the requirement that the AC must fill
T Average AC time at home, in years, between deployments
N Number of AC brigades that are fully ready, on average.

Formulas

$$s = a * f * (d/m) \tag{1}$$
$$R = B - s \tag{2}$$
$$T = [(A/R) - 1] * D \tag{3}$$
$$N = A * (T - P)/(T + D) \tag{4}$$

Derivation of Formulas

Formula (1) proceeds first from an expression for the number of brigades that can be mobilized at any time. That expression is the product of the total number available and the fraction of time that each can be mobilized:

$$a * f \tag{5}$$

However, during mobilization the unit can be deployed only a fraction of the mobilization period. That fraction is given by an expression equal to the length of each in-theater deployment period divided by the length of the mobilization period:

$$d/m \tag{6}$$

Therefore, the number of slots the RC can fill is the product of expressions (5) and (6): the number of units mobilized at any time multiplied by the fraction of time they are deployed. That product yields formula (1).

Formula (2) simply subtracts the RC slots from the total requirement.

Formula (3) proceeds from the observation that the ratio A/R represents the number of consecutive deployments in a full cycle in which all force elements are used. For example, if 12 AC brigades are available and three are required in each deployment, a complete cycle (using all available brigades) will last 4 deployment periods. After 4 periods, the units in the original phase of the cycle will deploy again, beginning a new cycle. In that case, A/R = 4.

Furthermore, each deployment period is D years long. Since the cycle lasts A/R periods, the full cycle length in years must be:

$$(A/R) * D \tag{7}$$

During the cycle, the unit will spend D years deployed. Therefore, its time at home must be:

$$[(A/R) * D] - D \tag{8}$$

That expression reduces to formula (3).

Formula (4) proceeds from the observation that a unit's total cycle may be represented as in Figure A.1.

Figure A.1
Representation of a Unit's Total Cycle

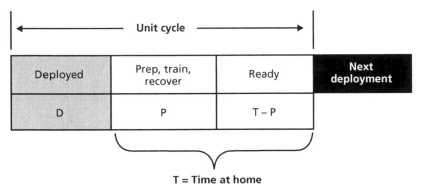

P represents time within the cycle when the unit is at home and either recovering from a previous deployment, conducting normal training on its mission-essential task list or wartime tasks, or preparing for the next deployment. In this report, the value of P is taken to be 11 months: two months for recovery, six months for training, and three months for predeployment preparation.

The total cycle time is given by the expression:

$$T + D \tag{9}$$

The amount of time ready is the amount of time at home minus the amount of time needed to recover, train up, and prepare for the next deployment:

$$T - P \tag{10}$$

Therefore the fraction of each brigade's total cycle time when it is ready is the expression:

$$(T - P)/(T + D) \tag{11}$$

At any given time, the proportion of all units that are ready must be equal to the fraction of the time when a brigade is ready—that is, expression (11). Therefore, the number of ready units at any given time is given by the expression:

$$A * (T - P)/(T + D), \text{ which is formula (4).}$$

Example

Consider the case in which 16 brigades are required, of which 11 must be heavy-medium; the AC has 41 transformed brigades available for rotation, of which 23 are heavy-medium; the RC has 25 heavy brigades. Focusing on filling the heavy/medium unit requirement, the assumptions for input parameters are therefore:

a = 25   Inventory includes 25 heavy-medium RC brigades available for rotation f = 1/6   RC brigades are to be mobilized, over the long term, no more than one out of six years

d = 0.5   Each RC brigade is deployed for six months (one-half year)

m = 1.0   Each RC brigade is mobilized for one year (six months of which is for preparation and recovery, while the other six months it is deployed)

B = 11   Each rotation requires 11 heavy-medium brigades

P = 11/12 Each AC unit has 11 months when it is not ready and available (two months recovery, six months train-up, and three months preparation for the next deployment)

A = 23   23 heavy-medium AC brigades are available for rotation

D = 1   Each AC deployment lasts one year.

The calculated values are:

$$s = a * f * (d/m) = 2.08$$
$$R = B - s = 8.92$$
$$T = [(A/R) - 1]*D = 1.58$$
$$N = A * (T - P)/(T + D) = 5.91$$

This implies the following conclusions:

- The RC can fill 2.08 slots in the requirement.
- The AC must fill the remaining 8.92 slots in the requirement.
- AC time at home between deployments is 1.58 years.
- The AC will have, on average, six ready AC brigades at any time.

Estimating Individual Time Away from Home

Authorization Structure for 19Ks

Individual time away from home over a career, as discussed in Chapter Three, was estimated separately for two groups of 19K soldiers: those in the junior segment of their career (grades E-4 and below) and those in the senior segment (grades E-5 and above). To conduct our calculations, we needed counts of soldiers by the junior-senior categories and also by type of unit (rotating TOE units, nonrotating TOE units (i.e., those in Korea), and TDA units.

Details of the 19K distribution have not yet been worked out for all of those groups, so we based our assessment on authorizations for the baseline force, adjusted for anticipated changes in heavy and medium units as the Army undergoes modular transformation to 43 transformed brigades. The baseline force is shown in Table A.3, drawn from Army authorization documents.

To adjust these figures for the transformed force, we assumed that one brigade in Korea (half of the TOE personnel) would be returned to the United States and would participate in the rotational schedule. In addition, we assumed that the planned conversion of divisions to the new structure would add 115 new 19Ks to the Army's requirements.[3] In total, those changes would require 9,183 19Ks in the new force, with fewer in Korea and more in the United States, as shown in Table A.4.

Table A.3
Number of 19K Authorizations by Grade Group and Unit Type (Baseline)

Grade Group	19K Authorizations, by Unit Type			
	TOE Units, U.S. and Europe	TOE Units, Korea	TDA Units	Total
Junior	3,810	334	298	4,442
Senior	2,777	269	1,581	4,627
Total	6,587	603	1,879	9,069

---

[3] These estimates reflect information as of late 2004 and could change. However, reasonable changes within the basic plan would not substantially affect our calculations. We assumed that the 3rd Infantry Division, 4th Infantry Division, and 1st Cavalry Division would convert from BCTs to UAs; each division's four UAs would then contain about 1,000 19Ks. Compared with their pretransformation authorizations (828 in 3rd Infantry Division and 950 and 967 in the others), that adds 255 19K positions. However, one of the brigades in 2nd Infantry Division will be converted to a light unit while another brigade will become a Stryker unit, which would subtract 141 19K positions. Thus, the net change is about 115 positions.

Table A.4
Distribution of 19K Authorizations by Grade Group and Unit Type
(Posttransformation)

| | Percentage of 19K Authorizations, by Unit Type | | | |
Grade Group	TOE Units, U.S. and Europe	TOE Units, Korea	TDA Units	Total
Junior	89.7	3.7	6.6	4,499
Senior	63.4	2.9	33.8	4,684
Total	76.2	3.3	20.5	9,183

Parameters and Formulas

From the above structure and other data on deployment cycles, we estimated values for the following inputs. Each was estimated separately for junior and senior career segments, but for simplicity we do not add subscripts here to distinguish the two segments.

Inputs (calculated for each career segment)

R Proportion of time spent in rotating elements of the force (estimated by the percentage of authorizations in the United States and Europe)

K Proportion of time spent in Korea (estimated by the percentage of authorizations in Korea)

d Proportion of time that rotating units are deployed (estimated from the unit cycle calculations described in the text)

f Proportion of time that rotating units spend in field training when they are not deployed (estimated from historical data).[4]

Calculated Values (calculated for each career segment)

D Proportion of segment deployed

F Proportion of segment in field training (while in a rotating unit and not deployed)

---

[4] We estimated the fraction f to be .23, based on deployment tempo reports on armor unit field training time before the onset of the war on terrorism, in 1999. That is, when armor units were not deployed, they spent about 23 percent of their time in the field for overnight training exercises. See Sortor and Polich (2001).

A Proportion of time away from home, considering all three sources.

Formulas

$$D = R * d \tag{12}$$
$$F = R * (1 - d) * f \tag{13}$$
$$A = D + F + K \tag{14}$$

Formula (12): The quantity R represents the fraction of service time that a soldier spends in a rotating TOE unit. The quantity d represents the fraction of time that each such unit spends deployed. Their product represents the proportion of all service time that was spent deployed. (Note: "service time" means the amount of time spent in the particular career segment being evaluated, either the junior or senior segment.)

Formula (13): The quantity R represents the fraction of service time that a soldier spends in a rotating TOE unit. The quantity (1 − d) represents the fraction of time that the unit spends at home station (i.e., not deployed). During that home-station period, the quantity f represents the fraction of time the unit spends in field training. The product of those three quantities represents the proportion of service time that an individual can expect to be in a rotating TOE unit and not on a deployment *and* away from home doing field training.

Formula (14): This one simply adds three proportions of service time for deployments, field training, and Korea.

Combination of Both Segments of Career

The above calculations produce two values for the parameter A, one for the junior and one for the senior segment of a career. We calculated a weighted average to combine both values into a single measure for a career, using weights of .25 for the junior segment and .75 for the senior segment.[5]

---

[5] Army information sources indicate that the typical soldier who remains through the first term remains in the Army, on average, for 16 years. We assumed that the first term occupies the first four years (typically in grades E-4 and below) and the remainder of a career is 12

Example

Consider the case of a 19K soldier in the *junior career segment*, under a unit cycle that calls for 16 total brigades deployed and using the transformed force with 23 AC heavy-medium transformed brigades and 11 RC heavy-medium transformed brigades. In that case, the unit cycle calculations show that time at home is 1.28 years. The unit also spends one year deployed during the cycle. Therefore the proportion of the unit's cycle time that is spent deployed is:

$$d = .439 \qquad \text{That is, } d = 1 - (1.28/2.28).$$

The other inputs for that case are the following:

R = .897 Fraction of junior 19K authorizations in U.S. and Europe TOE units

K = .037 Fraction of junior 19K authorizations in Korea TOE units

f = .23   Based on historical data for armor units, as explained above.

Applying the above formulas, we get these parameters for the junior segment of a career:

$$D = R * d = .897 * .439 = .394$$
$$F = R * (1 - d) * f = .897 * (1 - .439) * .23 = .116$$
$$A = D + F + K = .394 + .116 + .037 = .547.$$

This is the proportion of time away while in the junior segment of a career.

Similar calculations for the senior career segment yield a value of .388 for the proportion A. The proportion of time away for the entire career is the weighted average:

---

years (typically in grades E-5 and above). Thus the fraction of time spent in the junior grades is .25. Overall, the calculations are not very sensitive to this parameter.

Total time away during a career =
.25 * (.547) + .75 * (.388) = .428.

Hence, in this case, the typical 19K soldier spends 42.8 percent of his career away from home.

Estimating Costs

Here we describe the cost analysis that underlies the discussion of the long-term policy options in Chapter Four. That discussion presents seven options, A through G, which differ in the structure and utilization of AC and RC brigades. We attempted to assess the minimum costs associated with each option using the Army's FORCES model (March 2005 version) (FORCES is a limited-access cost-estimation tool), which enumerates detailed cost elements in such categories as direct equipment and parts, supplies and ammunition, personnel, indirect support, base operations, and acquisition of individual line items of equipment. (All costs in this report are in 2005 dollars.)

These estimates should be regarded as lower-bound estimates of cost. As detailed as the figures from the FORCES model are, they omit some cost elements that we will note below. In addition, we recognized some elements of cost that we cannot estimate at this time. We will point out both types of omitted cost elements when we suspect that they could lead to large additional costs in the future.

Mobilizing National Guard Brigades

In every option except Option A, reserve forces are mobilized and deployed in active-duty status to support rotational requirements. According to the FORCES model, the cost of operating a heavy transformed brigade on active duty will be $358 million a year. Because a National Guard unit mobilized for a year would incur the same personnel costs and similar training, we took this as a reasonable approximation for the annual costs of operating such a unit. These costs are offset slightly, however, because an activated National Guard unit would not conduct its normal training during that year, a savings of $62 million in military personnel and direct operating costs. Therefore, in our analysis of each of the options, we estimated the net cost of a mobilized National Guard transformed heavy brigade to be $296 million a year. Nevertheless, this estimate is clearly low: the operating cost from the model does not include any retirement accrual, health-care payments, installation-based subsidies, or veterans benefits that would be incurred.[6] Therefore, our estimates of operating the National Guard brigades in all of the options should be considered minimum costs.

Each of the seven options differs in the extent to which it uses National Guard brigades. Table A.5 shows our figures for the long-term average number of RC brigades that would be rotating each year and the resulting minimum annual operating costs (the product of $296 million times the number of rotating brigades).

Option B requires no start-up costs and no other annual costs beyond the mobilization of RC brigades. Therefore, its estimate is $272 million in annual costs. Similarly, Option F has no other costs to be considered, so its estimate is $841 million in annual costs.

Table A.5
RC Mobilization Costs

Option	Number of Brigades Rotating[a]	Minimum Annual Cost Estimate ($ Millions)
A	0.00	0
B	0.92	272
C	1.65	488
D	2.09	619
E	3.75	1,110
F	2.84	841
G	0.92	272

[a] This is parameter s in the formula defined earlier in this appendix.

---

[6] For example, the Congressional Budget Office estimates that the average annual cost for an AC soldier was about $99,000 in 2002, including cash and noncash benefits (CBO, 2004). In the Army FORCES model, the compensation portion for a transformed heavy brigade works out to only about half of that.

Training National Guard Brigades

To reduce the training time of National Guard brigades between mobilization and deployment, we increased the amount of unit training before mobilization in Options C and E. According to the FORCES model, a year's training for a National Guard transformed heavy brigade is $90 million, but only $62 million of that is for the direct costs that are likely to increase with more annual training—i.e., the costs for military personnel and for direct equipment, parts, and fuel. Prorated over four years, the time the unit is at home, the cost would be about $15 million a year.

For Option C, when 1.65 brigades are in the annual rotation, the cost would be $26 million. Adding that figure to the $488 million in annual mobilization costs (Table A.5) yields a total annual cost for Option C of $514 million.

Option E has on average 3.75 brigades in the annual rotation, and so our estimated cost is $58 million for the additional training. However, this is not the only other cost in Option E. We will return to that option below.

Beyond premobilization training, the Army would probably incur additional costs to improve postmobilization training by, for example, increasing the throughput of existing training facilities and ranges. We were unable to estimate the postmobilization costs, but they could be substantial. This is another way in which our cost estimates should be regarded as a lower bound.

Acquiring and Operating Transformed RC Heavy Brigades

Options D and E retain more heavy brigades in the National Guard, which would incur two other types of costs.

Start-Up Costs. First, the Army would have to convert 14 more RC brigades than planned to transformed heavy brigades. According to the FORCES model, the cost would be about $5.8 billion, which includes the costs for buying only the new equipment that would be needed to match the current unit with specifications. These conversion costs would be offset by avoiding the conversions of 14 heavy BCTs to transformed infantry brigades, a savings of about $2 billion.

Thus, the net start-up costs for Options D and E would be about $3.8 billion.

These one-time conversion costs to create transformed brigades could, however, be much larger than the model implies because many of the divisional brigades do not have a full complement of modern equipment and some contain less structure (e.g., only two maneuver battalions) than is called for the Army's new transformed brigades. For example, if none of the 14 converted brigades had modern equipment and the Army had to acquire a full set for each of them, the FORCES model indicates that the cost would be approximately $14 billion.

Annual Costs. Second, exercising Options D and E would incur an increase in the annual nonmobilized operating cost of the 14 transformed heavy brigades because these are more expensive to operate than transformed infantry brigades. According to the FORCES model, the direct operating costs and base support for a National Guard transformed heavy brigade would increase by more than $24 million per year relative to a transformed infantry brigade. This includes the increase in direct operating costs and base support, but not personnel. To cover all of the 14 new transformed heavy brigades, that cost would be about $341 million per year.

Again, these annual costs could be higher than estimated here for a variety of reasons. For example, many of the divisional National Guard brigades have historically been undermanned. To sustain full manning would require more robust recruitment and retention programs across all brigades, not just those that are mobilized at any one time—thus boosting annual costs across the force. Similar cost increases might be incurred in training and maintenance, which are considerably more expensive for heavy units than they are for light units, if heavy units were to incur higher costs than estimated by the model.

Adding the preceding cost elements yields the following totals for minimum annual costs. For Option D, we have $619 million for mobilization costs and $341 million for operating costs: a total of $960 million per year.

For Option E, we have $1.11 billion for mobilization costs, $58 million for increased RC premobilization training, and $341 million for operating costs: a total of more than $1.5 billion per year.

Acquiring and Operating Transformed AC Heavy Brigades

Option G increases by seven the number of AC transformed heavy brigades. This could be achieved either by adding new forces or converting infantry units into heavy units.

Start-Up Costs: Adding New Brigades. If the Army were to add new brigades to the force structure and increase its end strength in the process, it would incur three types of costs: equipping and training the new units, building base and training facilities, and manning the units. According to the FORCES model, the cost to add a new transformed heavy brigade in the AC would be $965 million. So for seven of these, our estimate of the cost is $6.8 billion to staff, equip, and train these units.

According to the Congressional Budget Office (2005), it could cost up to $525 million for each new transformed brigade to build new headquarters; operational facilities and infrastructure; morale, welfare, and recreation facilities; barracks; and schools for dependents. This would add about $3.7 billion in construction costs (although the figure could be lower if the Army were able to use existing barracks and facilities). We added this $3.7 billion to the preceding $6.8 billion figure, suggesting that total start-up costs could be at least $10 billion.

While these are significant costs, this estimate omits several important factors that are difficult to assess at this time but could increase the costs even further. For example, it does not include the costs to add support units that might be required to make the new transformed heavy brigades effective. Historically, Army support units have accounted for more soldiers than the combat units they support. That pattern, if replicated, could double the cost of additional manpower. It also excludes the operating costs for family housing for the new brigades as well as any accompanying training and other costs in the institutional Army.

Start-Up Costs: Converting Infantry Units. The costs would be lower if existing active component infantry units were instead turned into transformed heavy brigades. According to the FORCES model, the cost of converting seven existing infantry brigades to transformed heavy brigades would be $5.9 billion, offset by $700 million for not converting these seven infantry brigades to transformed infantry brigades. Thus, our estimate is that this would impose a start-up cost of $5.2 billion.

Operating Costs. In either approach, there would be the additional costs of operating the seven new transformed heavy brigades. Based on the Army FORCES estimate of $358 million per unit, the annual operating cost of operating seven new transformed brigades would be about $2.5 billion. If instead existing infantry units in the AC were converted, the net increase in their operating costs would be about $1.4 billion. In each case in this option, the Army would also incur at least $272 million in annual RC mobilization costs. Therefore, we estimate total annual costs to be at least $2.8 billion per year if new heavy brigades are formed, and about $1.7 billion if infantry units are converted to heavy units. Again, these costs could be higher than estimated if additional manpower were required for support units or for the institutional Army.

In sum, our minimum estimate for increasing AC force structure in Option G involves start-up costs ranging from about $5 billion to more than $10 billion and operating costs ranging from $1.7 billion to $2.8 billion.

Bibliography

Army National Guard, 81st Armor Brigade, *Deployment Overview and Timeline,* available at http://81brigade.washingtonarmyguard.com/Time line.html, accessed 2005.

Bowman, Tom, "Army Worries About Quality," *Baltimore Sun,* March 7, 2005, p. 1A.

Chu, David, Under Secretary of Defense for Personnel and Readiness, statement before the House Armed Services Committee, Washington, D.C., July 7, 2004.

Cody, Lieutenant General Richard A., hearing of the House Armed Services Committee, 108th Congress, Second Session, Washington, D.C., July 7, 2004.

_____, hearing of the Military Personnel Subcommittee of House Armed Services Committee, 109th Congress, First Session, Washington, D.C., February 2, 2005.

Cody, Lieutenant General Richard A., and Lieutenant General Franklin L. Hagenbeck, hearing of House Armed Services Subcommittee on the Total Force, Washington, D.C., March 10, 2004.

Congressional Budget Office, *An Analysis of the U.S. Military's Ability to Sustain an Occupation of Iraq,* Washington, D.C., 2003.

_____, *Military Compensation: Balancing Cash and Noncash Benefits,* Washington, D.C., 2004.

_____, *Options for Changing the Army's Overseas Basing,"* Washington, D.C., 2005.

Davey, Monica, "The New Military Life—Heading Back to the War," *New York Times,* December 20, 2004, p. 1.

Davis, Lynn E., David E. Mosher, Richard Brennan, Michael D. Greenberg, K. Scott McMahon, and Charles Woodruff Yost, *Army Forces for Homeland Security*, Santa Monica, Calif.: RAND Corporation, MG-221-A, 2004.

Department of Defense, *Report of the Quadrennial Defense Review,* Washington, D.C., 1997.

_____, *Quadrennial Defense Review Report*, Washington, D.C., 2001.

_____, *Rebalancing Forces: Easing the Stress on the Guard and Reserve*, Office of the Assistant Secretary of Defense for Reserve Affairs (Mobilization and Readiness), Washington, D.C., January 2004a.

_____, *Stability Operations Joint Operating Concept*, Draft, September 2004b, available at http://www.dtic.mil/jointvision/so_joc_v1.doc.

_____, *DoD Announces OEF/OIF Rotational Units*, Office of the Assistant Secretary of Defense (Public Affairs), news release, December 14, 2004c.

Department of the Army, *Unit Status Reporting*, Army Regulation 220-1, Washington, D.C., 1997.

_____, *The Army Strategic Planning Guidance 2006–2023*, Washington, D.C., 2003a.

_____, *Status of Force Rebalancing*, briefing by Office of the Deputy Chief of Staff for Operations and Plans, G-3, Washington, D.C., September 2003b.

_____, *Ground Troop Rotation Plan*, briefing slides, Office of the Deputy Chief of Staff for Operations and Plans, G-3, Washington, D.C., November 2003c, available at www.defenselink.mil/news/Nov2003/031106-D-6570C-001.pdf.

_____, *Army Campaign Plan*, Office of the Deputy Chief of Staff, G-3, Washington, D.C., February 7, 2004a, available at http://www.army.mil/thewayahead/acppresentations/4_12.html.

_____, *Building Army Capabilities*, briefing for media roundtable, Deputy Chief of Staff, G-3, Washington, D.C., February 17, 2004b.

_____, *2005 Army Modernization Plan*, February 2005a, available at http://www.army.mil/features/MODPlan/2005/.

_____, *2005 Army Posture Statement,* Washington, D.C., February 2005b, available at http://www.army.mil/aps/05/.

_____, *Army Strategic Planning Guidance 2005,* Washington, D.C., 2005c.

Feickert, Andrew, *U.S. Army's Modular Redesign: Issues for Congress,* Washington, D.C., Congressional Research Service, Library of Congress, RL32476, July 19, 2004, updated January 6, 2005.

Government Accountability Office (GAO), *Military Personnel: DOD Needs to Address Long-Term Reserve Force Availability and Related Mobilization and Demobilization Issues,* Washington, D.C., September 2004.

_____, *A Strategic Approach Is Needed to Address Long-Term Guard and Reserve Force Availability,* Washington, D.C., February 2, 2005.

Harvey, Francis J., Secretary of the Army, hearing of the House Armed Services Committee, Washington, D.C., February 9, 2005.

Hosek, James, and Mark Totten, *Serving Away from Home: How Deployments Influence Reenlistment,* Santa Monica, Calif: RAND Corporation, MR-1594-OSD, 2002.

Hunter, Duncan, Chairman, House Armed Services Committee, "House and Senate Conferees Approve 2005 Defense Authorization Act," press release, Washington, D.C., October 8, 2004.

Joint Chiefs of Staff, *National Military Strategy of the United States of America,* Washington, D.C., 2004.

Klein, Rick, "Soldiers Headed to Iraq Grill Rumsfeld: Blunt Queries Cite Safety, Long Tours," *Boston Globe,* December 9, 2004, p. A1.

Lippiatt, Thomas F., James C. Crowley, Patricia K. Dey, and Jerry M. Sollinger, *Postmobiliziation Training Resource Requirements: Army National Guard Heavy Enhanced Brigades,* Santa Monica, Calif.: RAND Corporation, MR-662-A, 1996.

Moniz, Dave, "For Guard Recruiters, a Tough Sell," *USA Today,* March 8, 2005.

Myers, General Richard B., Chairman of the Joint Chiefs of Staff, posture statement before the House Armed Services Committee, 109th Congress, February 16, 2005.

Office of Management and Budget (OMB), letter from the director to the Speaker of the House of Representatives, request for FY 2005 supplemental appropriations, February 14, 2005, available at http://www.whitehouse.gov/omb/budget/amendments/supplemental_2_14_05.pdf.

Office of the President, *The National Security Strategy of the United States of America*, Washington, D.C., September 2002.

Office of the Secretary of Defense (OSD), Budget Testimony, FY 2006, February 2005.

Pickup, Sharon, and Janet St. Laurent, *Force Structure: Preliminary Observations on Army Plans to Implement and Fund Modular Forces*, Government Accountability Office, GAO-05-443T, March 16, 2005, available at http://www.house.gov/hasc/testimony/109thcongress/Military%20Perso nnel/3-16-05GAOPickup.pdf.

President Bush's FY 2006 Defense Budget, available at http://www.dod.mil/news/Feb2005/d20050207budget.pdf.

Preston, Sergeant Major Kenneth, hearing testimony, Subcommittee on Military Quality of Life and Veterans Affairs, House Appropriations Committee, Washington, D.C., February 16, 2005.

Rumsfeld, Donald H., Secretary of Defense, *Rebalancing Forces*, memorandum for Secretaries of the Military Departments, Chairman of the Joint Chiefs of Staff, and Under Secretaries of Defense, Washington, D.C., July 9, 2003.

Schmitt, Eric, "Guard Reports Serious Drop in Enlistment," *New York Times,* December 17, 2004, p. 32.

Schoomaker, Peter J., Chief of Staff, U.S. Army, hearing before the House Armed Services Committee, Washington, D.C., November 17, 2004.

_____, Chief of Staff, U.S. Army, hearing before the Senate Armed Services Committee, 109th Congress, First Session, Washington, D.C., February 10, 2005.

Schultz, Lieutenant General Roger C., Chief, Army National Guard, hearing of Military Personnel Subcommittee, House Armed Services Committee, 109th Congress, First Session, February 2, 2005.

Schwartz, General Norton A., Director of Operations, Joint Staff, testimony before House Armed Services Committee, Washington, D.C., July 7, 2004.

Shanker, Thom, "Army Is Told to Plan for Shorter Tours in Iraq," *New York Times*, October 19, 2004, p. 23.

Sherman, Jason, "U.S. Revises Threat Scenarios," *Defense News,* November 22, 2004a.

_____, "Facing a New Reality: Nontraditional Threats Change Pentagon's Weapons Priorities," *Armed Forces Journal,* Vol. 143, No. 5, December 2004b, p. 22.

Sortor, Ronald E., Thomas F. Lippatt, J. Michael Polich, and James C. Crowley, *Training Readiness in the Army Reserve Components*, Santa Monica, Calif.: RAND Corporation, MR-474-A, 1994.

Sortor, Ronald E., and J. Michael Polich, *Army Deployments and Personnel Tempo*, Santa Monica, Calif.: RAND Corporation, MR-1417, 2001.

U.S. Congress, *National Defense Authorization Act for Fiscal Year 2000,* Public Law 106-65, Washington, D.C., 1999.

U.S. House of Representatives, Armed Services Committee, hearing on status of U.S. Armed Forces, November 17, 2004.

_____, hearing of the Military Personnel Subcommittee of the Armed Services Committee, 109th Congress, First Session, February 2, 2005a.

_____, hearing of the House Armed Services Committee, 109th Congress, First Session, February 9, 2005b.

U.S. Senate, Appropriations Committee, press release, June 22, 2004.

_____, hearing before the Armed Services Committee, March 3, 2005.